Springer Tracts in Modern Physics
Volume 153

Springer
Berlin
Heidelberg
New York
Barcelona
Hong Kong
London
Milan
Paris
Singapore
Tokyo

Springer Tracts in Modern Physics

Springer Tracts in Modern Physics provides comprehensive and critical reviews of topics of current interest in physics. The following fields are emphasized: elementary particle physics, solid-state physics, complex systems, and fundamental astrophysics.

Suitable reviews of other fields can also be accepted. The editors encourage prospective authors to correspond with them in advance of submitting an article. For reviews of topics belonging to the above mentioned fields, they should address the responsible editor, otherwise the managing editor.

See also http://www.springer.de/phys/books/stmp.html

Managing Editor

Gerhard Höhler

Institut für Theoretische Teilchenphysik
Universität Karlsruhe
Postfach 69 80
D-76128 Karlsruhe, Germany
Phone: +49 (7 21) 6 08 33 75
Fax: +49 (7 21) 37 07 26
Email: gerhard.hoehler@physik.uni-karlsruhe.de
http://www-ttp.physik.uni-karlsruhe.de/

Elementary Particle Physics, Editors

Johann H. Kühn

Institut für Theoretische Teilchenphysik
Universität Karlsruhe
Postfach 69 80
D-76128 Karlsruhe, Germany
Phone: +49 (7 21) 6 08 33 72
Fax: +49 (7 21) 37 07 26
Email: johann.kuehn@physik.uni-karlsruhe.de
http://www-ttp.physik.uni-karlsruhe.de/~jk

Thomas Müller

Institut für Experimentelle Kernphysik
Fakultät für Physik
Universität Karlsruhe
Postfach 69 80
D-76128 Karlsruhe, Germany
Phone: +49 (7 21) 6 08 35 24
Fax: +49 (7 21) 6 07 26 21
Email: thomas.muller@physik.uni-karlsruhe.de
http://www-ekp.physik.uni-karlsruhe.de

Roberto Peccei

Department of Physics
University of California, Los Angeles
405 Hilgard Avenue
Los Angeles, CA 90024-1547, USA
Phone: +1 310 825 1042
Fax: +1 310 825 9368
Email: peccei@physics.ucla.edu
http://www.physics.ucla.edu/faculty/ladder/
peccei.html

Solid-State Physics, Editor

Peter Wölfle

Institut für Theorie der Kondensierten Materie
Universität Karlsruhe
Postfach 69 80
D-76128 Karlsruhe, Germany
Phone: +49 (7 21) 6 08 35 90
Fax: +49 (7 21) 69 81 50
Email: woelfle@tkm.physik.uni-karlsruhe.de
http://www-tkm.physik.uni-karlsruhe.de

Complex Systems, Editor

Frank Steiner

Abteilung Theoretische Physik
Universität Ulm
Albert-Einstein-Allee 11
D-89069 Ulm, Germany
Phone: +49 (7 31) 5 02 29 10
Fax: +49 (7 31) 5 02 29 24
Email: steiner@physik.uni-ulm.de
http://www.physik.uni-ulm.de/theo/theophys.html

Fundamental Astrophysics, Editor

Joachim Trümper

Max-Planck-Institut für Extraterrestrische Physik
Postfach 16 03
D-85740 Garching, Germany
Phone: +49 (89) 32 99 35 59
Fax: +49 (89) 32 99 35 69
Email: jtrumper@mpe-garching.mpg.de
http://www.mpe-garching.mpg.de/index.html

Radoje Belušević

Neutral Kaons

With 67 Figures

Springer

Dr. Radoje Belušević

High Energy Accelerator Research Organization KEK
Department of Physics
1-1 Oho, Tsukuba-shi
305-0801 Ibaraki-ken, Japan
Email: belusev@cccemail.kek.jp

Physics and Astronomy Classification Scheme (PACS): 14.40.A, 03.65.-w, 13.20.Eb, 13.25.Es, 11.30.Er, 12.15.Ff, 12.15.Ji

ISSN 0081-3869
ISBN 3-540-65645-6 Springer-Verlag Berlin Heidelberg New York

Library of Congress Cataloging-in-Publication Data applied for.

Die Deutsche Bibliothek – CIP Einheitsaufnahme

Belušević, Radoje: Neutral kaons/Radoje Belušević. – Berlin; Heidelberg; New York; Barcelona; Hong Kong; London; Milan; Paris; Singapore; Tokyo: Springer, 1999
(Springer tracts in modern physics; Vol. 153)
ISBN 3-540-65645-6

Typesetting: Data conversion by EDV-Beratung F. Herweg, Hirschberg
Cover design: *design & production* GmbH, Heidelberg

SPIN: 10709232   56/3144 - 5 4 3 2 1 0 – Printed on acid-free paper

Dedicated to **Jack Steinberger**

Foreword

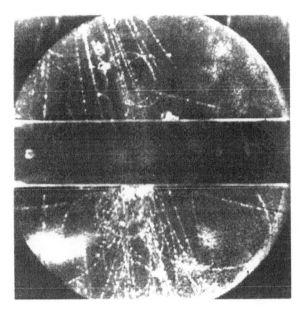

Among the five thousand stereoscopic photographs of cosmic ray showers obtained by George Rochester and Clifford Butler at Manchester University, using a cloud chamber placed in a magnetic field, there was a picture containig "forked tracks of a very striking character". In the lower right-hand side of the picture, just below a 3-cm lead plate mounted across the centre of the chamber, they observed, on 15th October 1946, a pair of tracks forming a two-pronged fork (an inverted V) with the apex in the gas (see the reprinted image).

The direction of the magnetic field was such that a positively charged particle moving downward is deflected in an anticlockwise direction. They determined that the particle corresponding to the upper track had positive charge and a momentum of 340 ± 100 MeV/c; the lower particle had negative charge and a momentum of 350 ± 150 MeV/c. The ionization and curvature of the tracks showed that they were due to particles much less massive than the proton.

If the tracks were associated with a collision process, one would have expected several hundred times as many of these interactions in the lead plate as in the gas. Since very few events similar to this were observed in the plate, they argued that the fork "must be due to some type of spontaneous process for which the probability depends on the distance travelled and not on the amount of matter traversed". This conclusion is supported by the following argument: if the fork were due to a deflection of a backscattered charged particle by a nucleus, the momentum transfer would be so large as to produce a visible recoiling nucleus at the apex.

Based on their past experience, the electron pair production by a high-energy photon in the Coulomb field of the nucleus was excluded because the two tracks would have to be much closer together if they were an electron–positron pair. They also excluded the possibility of this picture representing the decay of a charged pion or muon coming up from below the chamber, since in that case conservation of energy and momentum would require the incident particle to have a minimum mass of $1280 m_e$ (m_e is the electron mass).

Rochester and Butler therefore concluded that this had to be a photographic image of the decay of a new type of uncharged elementary particle into two lighter charged particles. For the case where the incident particle decays into two particles of equal mass, they determined the mass of the parent particle to be 870 ± 200 MeV/c^2, for an assumed secondary particle mass of $200 m_e$.

Preface

Enormous progress has been made in the field of high-energy, or elementary particle, physics over the past three decades. The existence of a subnuclear world of quarks and leptons, whose dynamics can be described by quantum field theories possesing local gauge symmetry (*gauge theories*), has been firmly established. The cosmological and astrophysical implications of experimental results and theoretical ideas from particle physics have become essential to our understanding of the formation of the universe. For example, a tiny violation of CP symmetry, which has been observed so far only in the K^0 system, is believed to have played an important role in the early stages of cosmic evolution.

The main purpose of this book is to convey the unique beauty of a quantum-mechanical system that contains so many of the aspects of modern physics. Inevitably, this imposes considerable constraints on the content and nature of the presentation. In outlining the basic formalism necessary to describe the K^0 system and its time evolution in both vacuum and matter, effort was made to keep the presentation as clear as possible and to justify the main steps in the derivations. To highlight their quantum-mechanical origin, extraordinary properties of neutral kaons are illustrated through analogous experiments with polarized light and atomic beams. A formal theory of the discrete symmetry operations C (charge conjugation), P (parity transformation) and T (time reversal) is presented. These subtle concepts are discussed in the context of parity violation, time reversal asymmetry and CP noninvariance in kaon decays. In order to emphasize the complementary roles of theory and measurement, a number of "classic" experiments with neutral K mesons are described and some major current projects and proposals are reviewed. A detailed and pedagogical discussion of the K^0 physics within the framework of gauge theories of the electroweak interactions is also provided.

Athough this book was written primarily for graduate students and researchers in high-energy physics, I have endeavored to make its content accessible to curious undergraduates and physicists not specializing in the field.

Acknowledgements

I would like to thank Bruce Winstein and Italo Mannelli for valuable comments regarding the experiments E731 at Fermilab and NA31 at CERN.

I have benefitted from discussions with Robert Sachs about the K^0 phenomenology in the presence of T and CPT violation, and with Kaoru Hagiwara, Makoto Kobayashi, Yasuhiro Okada and Yasuhiro Shimizu concerning $K^0\text{-}\bar{K}^0$ mixing and rare kaon decays in the Standard Model. Helpful comments and suggestions by Volker Hepp, Martin Wunsch and Sher Alam are appreciated. I am particularly indebted to Asish Satpathy and Bruce Winstein for their interest, help and advice.

For permission to reprint various plots and drawings I am grateful to Bill Carithers, Val Fitch, Erwin Gabathuler, Jack Ritchie, Jack Steinberger and Bruce Winstein.

I wish to express my special gratitude to Hans Kölsch, Victoria Wicks and the production team at Springer for their help in preparing the manuscript for publication.

Support from Prof. Sakue Yamada, Head of the Institute for Particle and Nuclear Studies at KEK, and the Japanese Ministry of Education, Science and Culture (Monbusho) is gratefully acknowledged.

Tsukuba-shi   *R. Belušević*
February, 1999

Contents

1. Introduction

"This is one of the greatest achievements of theoretical physics. It is not based on an elegant mathematical hocus-pocus such as the general theory of relativity yet the predictions are just as important as, say, the prediction of positrons."

Richard Feynman, The Theory of Fundamental Processes

The neutral K meson (*neutral kaon*), K^0, and its antiparticle, $\bar{K}^0$, form a remarkable quantum-mechanical two-state system that has played an important role in the history of elementary particle physics. Indeed, ever since the discovery of K^0 half a century ago, neutral kaons have been a rich source of unique and fascinating phenomena associated with their production, decay and propagation in both vacuum and matter.

What makes the K^0 system so special is that K^0 and $\bar{K}^0$, which have the same charge, mass, spin and parity, but different *strangeness* quantum number, S, cannot always be distinguished from one another. Whereas in strangeness-conserving strong interactions K^0 ($S = +1$) and $\bar{K}^0$ ($S = -1$) are as distinct as the neutron and antineutron, this distinction is erased in strangeness-violating weak interactions, thus allowing $K^0 \leftrightarrow \bar{K}^0$ transitions.[1]

As a consequence, an initially pure $|K^0\rangle$ or $|\bar{K}^0\rangle$ state will gradually evolve into a state of mixed strangeness,

$$|K^0(t)\rangle \longrightarrow a(t)|K^0\rangle + b(t)|\bar{K}^0\rangle,$$

in accordance with the principle of superposition of amplitudes in quantum mechanics. This *strangeness oscillation* effect has a nice optical analogy: right-circularly polarized light rapidly acquires a large left-circularly polarized component while passing through a crystal that absorbs predominantly x-polarized light.

The K^0 and $\bar{K}^0$ mesons are two unconnected, degenerate ($m_{k^0} = m_{\bar{k}^0}$) states in the absence of the weak interaction. As is well known from quantum mechanics, the mixing of two degenerate levels in vacuum must result in level splitting (this splitting shows up in the hydrogen molecular ion and in the inversion spectrum of ammonia). The application of ordinary perturbation theory to the K^0-$\bar{K}^0$ system produces the following result: the weak interaction, $\hat{H}_w$, slightly shifts the value of the kaon mass, m_{k^0}, and splits the degenerate levels by a tiny amount:

$$\Delta m_k \equiv |m_1 - m_2| = \langle K^0|\hat{H}_w|\bar{K}^0\rangle + \langle \bar{K}^0|\hat{H}_w|K^0\rangle,$$

where Δm_k is the mass difference of the two states, K_1^0 and K_2^0, created by strangeness-changing $K^0 \leftrightarrow \bar{K}^0$ transitions. These new states are the correct

---

[1] The neutron and antineutron do not mix because of baryon number conservation.

linear superpositions of K^0 and $\bar{K}^0$, which diagonalize the perturbation. In turn, K^0 and $\bar{K}^0$ are linear superpositions of K_1^0 and K_2^0.

Since both K^0 and its antiparticle decay to two or more pions, the initial state in a pionic decay of a neutral kaon must be some linear combination of K^0 and $\bar{K}^0$ states. To see what this implies, suppose that weak interactions do not make an arbitrary distinction between particles and antiparticles, i.e., that they are invariant under the combined operation of charge conjugation and parity transformation (space inversion), $\hat{C}\hat{P}$. The wavefunction of a 2π final state does not change its sign under $\hat{C}\hat{P}$ (it is "even" under this symmetry transformation). Consequently, one linear combination of neutral kaon states, $K_1^0 \equiv K^0 + \bar{K}^0$, can decay into a pair of pions with no CP violation. The other, equally probable, combination $K_2^0 \equiv K^0 - \bar{K}^0$ cannot, because it is "odd" under $\hat{C}\hat{P}$. This state is thus forced to find other CP-conserving ways to decay, such as into three pions, in which case the relatively small three-body phase space makes the lifetime of K_2^0 much longer than that of K_1^0. This remarkable prediction was made in 1955 by M. Gell-Mann and A. Pais.

The unique beauty of the K^0 system stems from the quantum-mechanical interplay between the two related sets of particles

$$\left(K_1^0, K_2^0 \right) \quad \leftrightarrow \quad \left(K^0, \bar{K}^0 \right),$$

as illustrated by the following examples.

- The time evolution of the K^0-$\bar{K}^0$ system in vacuum (*the strangeness oscillation*) can be described in terms of the two CP eigenstates K_1^0 and K_2^0 which have different masses and lifetimes.
- Another spectacular quantum-mechanical phenomenon occurs because the K^0 and $\bar{K}^0$ mesons, by virtue of their opposite strangeness, interact quite differently with matter. Passing a pure K_2^0 beam, which is an equal mixture of K^0 and $\bar{K}^0$ particles, through a slab of material will alter the beam composition, resulting in a new linear combination of K_1^0 and K_2^0 states (*the K_1^0 regeneration*).
- The existence of the long-lived neutral kaon and the smallness of the K_1^0-K_2^0 mass difference indicate that the gravitational couplings of matter and antimatter are equal: if the gravitational potential energy of the K^0 meson were opposite to that of its antiparticle, the two would mix so rapidly that the K_2^0 meson could never be detected (M. L. Good, 1961).

The importance of the K^0 system transcends its quantum-mechanical intricacy. Many of the theoretical ideas that form the basis of our current understanding of particle physics are intimately associated with K mesons, in particular with neutral kaons.

In the early 1950s, the multitude of observed K-meson decays led to the realization that parity may be violated in weak interactions. Central to this development was a brilliant analysis by R. Dalitz who showed, in 1953, that the 2π and 3π decay modes of the charged kaon required the parent particles to have opposite intrinsic parities. This created a serious dilema: either the $K \to 2\pi$ and $K \to 3\pi$ decays were due to different initial states, or parity was not conserved. In 1956 T. D. Lee and C. N. Yang questioned the experimental basis for the assumption of parity conservation in weak interactions. Soon thereafter experiments suggested by them observed that parity was indeed violated. The dilema was thus resolved: the parent particles were the same, but parity was not conserved in the decays. The prevailing sentiment among physicists prior to this discovery was quaintly expressed by W. Pauli: "*What God hath put asunder no man shall ever join.*"

Subsequent investigations showed that parity violation was compensated by a failure of charge conjugation. This possibility was first suggested by L. Landau in 1957 before the observation of parity noninvariance! CP was consequently considered to be an exact symmetry of nature until "these same particles, in effect, dropped the other shoe". In 1964, J. Cronin and V. Fitch with their collaborators detected one 2π event among 500 or so common decays of the long-lived neutral kaon. This tiny violation of *CP* symmetry, which has been observed so far only in the K^0 system, remains a great mystery to this day, especially because it may have played an essential role in the early formation of the universe.

The concept of *strangeness*, introduced in 1953 by M. Gell-Mann and, independently, T. Nakano and K. Nishijima to explain the anomalously long lifetimes of K-mesons and hyperons, was crucial for the development of the *quark model* of particles (M. Gell-Mann, 1964). Quark *flavors*, such as strangeness, are conserved in strong interactions but not in weak decays, resulting in the long lifetimes of strange particles.

The smallness of the observed branching ratio for the decay $K_2^0 \to \mu^+\mu^-$, implying the absence of strangeness-changing neutral weak currents, led to the prediction of a fourth quark, the *charm* quark. This prediction is based on *quark mixing*, which lies at the heart of the assumed universality of the weak interactions of quarks and *leptons*, and thus of the highly successful *Standard Model* of elementary particles.

CP violation was introduced in the Standard Model by increasing the number of quark and lepton *families* to at least three (M. Kobayashi and T. Maskawa, 1973). This idea became very attractve with the subsequent discovery of the *bottom* quark, which forms, together with the recently detected *top* quark, a third family of quarks.

1.1 K^0 and $\bar{K}^0$ as Eigenstates of Strangeness

The neutral kaon[2] was discovered in 1946 by G. Rochester and C. Butler [1] in a cloud chamber exposed to cosmic rays. They observed a pair of charged

---

[2] Kaons behave in some respects like heavy pions and so they are included in the family of mesons.

particles, later identified as pions, that could be associated with the decay of a neutral particle about 900 times heavier than the electron.[3]

With the benefit of hindsight we can say that the particle they detected was either a K^0 or a $\bar{K}^0$. But how do we know that there is an antiparticle to the neutral kaon, i.e., how can we distinguish a K^0 from a $\bar{K}^0$? The answer is provided by the empirical fact that some strong-interaction processes, e.g.,

$$\pi^- + p \to K^0 + n, \quad K^- + p \to K^0 + \Lambda^0, \tag{1.1}$$

have not been observed, although they violate neither the charge nor the baryon number conservation. This prompted M. Gell-Mann and, independently from him, T. Nakano and K. Nishijima to introduce, in 1953, a new quantum number called *strangeness* [2], which is conserved in strong but not in weak interactions. If we assign the quantum numbers I (isospin) and S (strangeness) according to Table 1.1 to the particles participating in (1.1), then those reactions are clearly not allowed. On the other hand, the strong interactions

$$K^+ + n \to K^0 + p, \quad \pi^- + p \to \Lambda^0 + K^0, \quad K^- + p \to \bar{K}^0 + n \tag{1.2}$$

are allowed and have indeed been observed. From (1.2) we see that K^+ is associated with K^0 and K^- with $\bar{K}^0$, thus forming two K doublets (see Table 1.1).

Table 1.1. Isospin and strangeness assignments.

				I_3		
S	I	-1	$-\frac{1}{2}$	0	$+\frac{1}{2}$	$+1$
0	$\frac{1}{2}$		n		p	
-1	0			Λ		
0	1	π^-		π^0		π^+
$+1$	$\frac{1}{2}$		K^0		K^+	
-1	$\frac{1}{2}$		K^-		$\bar{K}^0$	
-1	1	Σ^-		Σ^0		Σ^+
-2	$\frac{1}{2}$		Ξ^-		Ξ^0	
-3	0			Ω^-		
0	0			η		

---

[3] The K^0 was discovered in 1946 and the $K^\pm$ in 1947, both by G. Rochester and C. Butler. The discovery of $\pi^\pm$ by C. Lattes, G. Occhialini and C. Powell was reported a day after the first observation of the $K^\pm$! The first direct detection of the π^0 was made in 1950 by J. Steinberger, W. Panofsky and J. Steller.

Now, to distinguish a K^0 from a $\bar{K}^0$ we just have to let them interact with matter. For example, a $\bar{K}^0$ can produce a Λ particle (the so-called *associated production*),

$$\bar{K}^0 + p \to \Lambda^0 + \pi^+, \tag{1.3}$$

whereas a K^0 cannot. Therefore,

$$K^0 \neq \bar{K}^0 \text{ unlike } \gamma, \ \pi^0, \ \eta^0, \ \omega^0, \text{ etc.}$$

As already mentioned, strangeness is not conserved in weak interactions. As a consequence, K^0 and $\bar{K}^0$ can decay weakly into the same final states. If we observe their common decay modes only, the two neutral kaons look like the same particle; they can be distinguished only through their production in strong interactions.

The fact that K^0 and $\bar{K}^0$ have common decay modes,

$$K^0(\bar{K}^0) \to \pi^+\pi^-, \ \pi^0\pi^0, \ \pi^+\pi^-\pi^0, \ 3\pi^0, \text{ etc.} \tag{1.4}$$

suggests that they are connected through strangeness-violating ($|\Delta S| = 2$), second-order weak transitions (see Fig. 1.1). Therefore, the particles we observe in experiments are not K^0 and $\bar{K}^0$, but rather a linear superposition of the two. This situation is (almost) unique to the K^0 system (only $D^0\bar{D}^0$ and $B^0\bar{B}^0$ share this property), because the only quantum number that distinguishes them, S, is not conserved in weak interactions. This is not the case for other particle–antiparticle pairs, since their quantum numbers (charge, lepton number, baryon number) are conserved in all interactions.

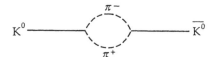

K^0 $\bar{K}^0$

Fig. 1.1. The K^0-$\bar{K}^0$ transition via an intermediate $\pi^+\pi^-$ pair

Describing the K^0 system as a linear superposition of K^0 and $\bar{K}^0$ states is analogous to representing photons as a linear superposition of right- and left-circularly polarized states. We shall thus express the states of neutral kaons by two-dimensional (complex) vectors. Let $|K^0\rangle$ denote the state in which the meson is a K^0, and $|\bar{K}^0\rangle$ the state in which it is a $\bar{K}^0$. The general state of a neutral kaon is then given by

$$|\Psi\rangle = a|K^0\rangle + b|\bar{K}^0\rangle, \quad |a|^2 + |b|^2 = 1. \tag{1.5}$$

The strangeness operator, $\hat{S}$, is defined by

$$\hat{S}|K^0\rangle = +|K^0\rangle, \quad \hat{S}|\bar{K}^0\rangle = -|\bar{K}^0\rangle. \tag{1.6}$$

1.2 CP Eigenstates of Neutral Kaons: K_1^0 and K_2^0

Weak interactions are not invariant under space inversion (parity transformation) $\hat{P}$. Indeed, all neutrinos have negative helicity (they are said to be left-handed) and all antineutrinos have positive helicity (they are right-handed), and so in this sense parity is maximally violated in weak interactions. As a consequence, charge conjugation $\hat{C}$ is also not conserved in weak interactions: applying $\hat{C}$ to a left-handed neutrino turns it into a left-handed antineutrino, a state which does not exist. In this section it will be assumed, however, that the combined operation of charge conjugation and parity transformation, $\hat{C}\hat{P}$, is conserved in weak interactions.[4]

We define the effect of $\hat{C}\hat{P}$ on K^0 and $\bar{K}^0$ to be[5]

$$\hat{C}\hat{P}|K^0\rangle = |\bar{K}^0\rangle, \quad \hat{C}\hat{P}|\bar{K}^0\rangle = |K^0\rangle, \tag{1.7}$$

i.e.,

$$|K^0\rangle, \ |\bar{K}^0\rangle \neq CP \text{ eigenstates}$$

In accordance with the above assumption, we form the following linear, orthonormal combinations of $|K^0\rangle$ and $|\bar{K}^0\rangle$, which are CP eigenstates:

$$|K_1^0\rangle \equiv \frac{1}{\sqrt{2}}\left[|K^0\rangle + |\bar{K}^0\rangle\right], \quad \hat{C}\hat{P}|K_1^0\rangle = +|K_1^0\rangle,$$

$$|K_2^0\rangle \equiv \frac{1}{\sqrt{2}}\left[|K^0\rangle - |\bar{K}^0\rangle\right], \quad \hat{C}\hat{P}|K_2^0\rangle = -|K_2^0\rangle. \tag{1.8}$$

Thus by assuming that CP is conserved in weak interactions, we infer that there is a particle that can decay only into a state with $CP = +1$ (K_1^0) and a particle which decays only into a $CP = -1$ state (K_2^0). According to (1.4), neutral kaons typically decay into two or three pions, i.e. into states with parity $+1$ and -1, respectively (both states have $C = +1$; see Appendix A). Therefore, K_1^0 is expected to decay only into two pions, whereas K_2^0 must decay only into three pions:

$$K_1^0 \to 2\pi, \quad K_2^0 \to 3\pi. \tag{1.9}$$

In the 2π decay, there is 215 MeV of kinetic energy available for the pions ($m_k \approx 500$ MeV, $m_\pi \approx 140$ MeV); in the 3π decay only 78 MeV. The small 3π phase space makes the lifetime of K_2^0 much longer than that of K_1^0 (see (1.12) and [21]).

---

[4] C and P are strictly conserved in strong and electromagnetic interactions. CP violation in weak decays of neutral kaons will be discussed later on.

[5] There is an arbitrary phase associated with this definition:

$$\hat{C}\hat{P}|K^0\rangle \equiv e^{i\theta_{\mathrm{cp}}}|\bar{K}^0\rangle \ \to \ \hat{C}\hat{P}|\bar{K}^0\rangle = \hat{C}\hat{P}\left(e^{-i\theta_{\mathrm{cp}}}\hat{C}\hat{P}|K^0\rangle\right) = e^{-i\theta_{\mathrm{cp}}}|K^0\rangle.$$

We choose $\theta_{cp} = 0$. The relative phase of K^0 and $\bar{K}^0$ is not fixed by the strong and electromagnetic interactions, since these states do not couple except through very feeble second-order weak transitions.

We can invert equations (1.8) to express $|K^0\rangle$ and $|\bar{K}^0\rangle$ as linear super-positions of $|K_1^0\rangle$ and $|K_2^0\rangle$ states:

$$|K^0\rangle \equiv \frac{1}{\sqrt{2}}\left[|K_1^0\rangle + |K_2^0\rangle\right],$$
$$|\bar{K}^0\rangle \equiv \frac{1}{\sqrt{2}}\left[|K_1^0\rangle - |K_2^0\rangle\right]. \tag{1.10}$$

Now, if a beam of K^0 particles is produced at $t = 0$, then at a later time $t \gg \tau_1$ (τ_1 is the lifetime of K_1^0) only K_2^0 mesons will survive:

$$|K^0\rangle \propto |K_1^0\rangle + |K_2^0\rangle \to |K_2^0\rangle, \quad t \gg \tau_1. \tag{1.11}$$

Near the beam production point one would thus see mainly 2π decays, and farther down the beamline only 3π decays.

There is a nice optical analogy to (1.11): if we were to shine a linearly polarized beam of light on a crystal that absorbs preferentially left-circularly polarized light, the beam would become increasingly right-circularly polarized as it passes through the crystal, just as a K^0 beam decays into a K_2^0 beam.

The remarkable prediction (1.11) was originally made in 1955 by M. Gell-Mann and A. Pais in a paper famed for the sheer beauty of their reasoning [3].[6] The first experimental confirmation was swift: in 1956, L. Lederman and his collaborators discovered the K_2^0 meson at Brookhaven in a cloud chamber[7] placed sufficiently far from the beam production point to allow all K_1^0 mesons and Λ^0 particles to decay before reaching the chamber [4] (see Fig. 1.2). Subsequent experiments, which could detect both the $\pi^+\pi^-$ and the $2\pi^0$ decays, showed that about a half of all originally produced K^0 mesons decayed by these two modes.

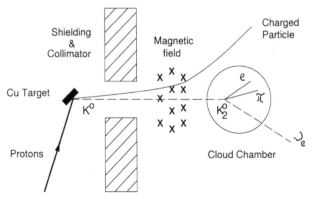

Fig. 1.2. The experimental setup used by K. Lande et al. [4, 5]

---

[6] This work was published before the discovery of parity violation. Nevertheless, the essence of their argument remains the same when $\hat{C}$ is replaced by $\hat{C}\hat{P}$.

[7] This was one of the last cloud chamber experiments.

The measured lifetimes of the two CP eigenstates K_1^0 and K_2^0 are

$$\tau_1 = 0.89 \times 10^{-10} \text{ s}, \quad c\tau_1 = 2.7 \text{ cm}$$
$$\tau_2 = 5.17 \times 10^{-8} \text{ s}, \quad c\tau_2 = 15.5 \text{ m}$$

$$(1.12)$$

The data for K^0 and $\bar{K}^0$ decays are mutually consistent in the sense that the lifetimes and decay modes for the K_1^0 and K_2^0 components are the same.

1.3 Duality of Neutral Kaons: $(K^0, \bar{K}^0)$ vs. (K_1^0, K_2^0)

Unlike K^0 and $\bar{K}^0$, the CP eigenstates K_1^0 and K_2^0 are not a particle–antiparticle pair: each is its own antiparticle and each has a unique lifetime; hence their masses may also differ. In contrast, K^0 and $\bar{K}^0$ have the same mass but do not have a unique lifetime, for they decay faster into two pions than into three pions.

According to (1.8) and (1.10),

- $K^0(\bar{K}^0)$ is an equal mixture of K_1^0 and K_2^0
- $K_1^0(K_2^0)$ is an equal mixture of K^0 and $\bar{K}^0$

The K^0 system, therefore, is described by two related sets of particles:

$$(K^0, \bar{K}^0) \longleftrightarrow (K_1^0, K_2^0).$$

Which set of particles is observed depends on the nature of the measurement: K_1^0, K_2^0 are detected through their decays into states with definite CP parity (2π or 3π), and $K^0, \bar{K}^0$ through mutually distinct semileptonic decays $K^0 \to e^+\pi^-\nu_e$ and $\bar{K}^0 \to e^-\pi^+\bar{\nu}_e$, which obey the empirical rule ΔStrangeness $= \Delta$Charge.

To summarize:

$K^0, \bar{K}^0$ $\begin{cases} \bullet \textit{ are distinguished through production (in strong} \\ \textit{ interactions) and through semileptonic weak decays,} \\ \bullet \textit{ are eigenstates of strangeness,} \\ \bullet \textit{ have no unique lifetime,} \\ \bullet \textit{ have the same mass,} \\ \bullet \textit{ are linear superpositions of } K_1^0 \textit{ and } K_2^0; \end{cases}$

K_1^0, K_2^0 $\begin{cases} \bullet \textit{ are distinguished through pionic weak decays,} \\ \bullet \textit{ are eigenstates of } CP, \\ \bullet \textit{ each has a unique lifetime,} \\ \bullet \textit{ have different mass,} \\ \bullet \textit{ are linear superpositions of } K^0 \textit{ and } \bar{K}^0. \end{cases}$

1.4 The Einstein–Podolsky–Rosen Paradox in the K^0 System

The decays of two-kaon states provide a unique insight into the dynamic behavior of quantum systems over macroscopic distances, and thus into the difference between classical and quantum concepts of measurement. This difference lies at the heart of the famous Einstein–Podolsky–Rosen "paradox".

The classical concept is outlined in the following remark by Albert Einstein: "The real factual situation of the system S_2 is independent of what is done with the system S_1, which is spatially separated from the former." This is in sharp contrast with the quantum-mechanical interpretation, according to which a measurement on what appears to be a part of the system is in fact a measurement on the whole system.

To understand the meaning of the latter statement, suppose that two neutral kaons are emitted in the $+z$ and $-z$ directions from the decay at rest of an odd eigenstate of $\hat{C}$ (the Φ meson):

$$|\Phi\rangle \longrightarrow |K^0\bar{K}^0\rangle, \quad J_\phi^{PC} = 1^{--}.$$

The two kaons emitted simultaneously are always in opposite states: one is K^0 and the other $\bar{K}^0$, or one is K_1^0 and the other K_2^0. The reason is that the initial state has odd parity ($P = -1$), while Bose statistics forbids odd parity states for two identical spinless bosons. Note also that Φ has zero strangeness, and that strangeness is conserved in the strong decay $\Phi \to K^0\bar{K}^0$.

Now, consider a coincidence measurement of the decays of kaons emitted in opposite directions. If we detect the pionic decay $K_1^0 \to 2\pi$ of the kaon emitted in the $+z$ direction, then its companion in the $-z$ direction must be a K_2^0. However, if we detect the semileptonic decay $K^0 \to e^+\pi^-\nu_e$ at $+z$, then the kaon at $-z$ is definitely a $\bar{K}^0$. Even when the two kaons are very far apart with no possibility of interacting, we can still "select" the state of the kaon at $-z$ based on what we choose to measure at $+z$.

In quantum mechanics, all relevant information about a system is contained in its wave function. When two systems have wave functions which differ at most by a constant phase factor, they are considered to be in the same quantum state, which is a linear superposition of the individual state vectors. We can thus write

$$|\Phi\rangle = \frac{1}{\sqrt{2}} \left\{ |K^0(z)\bar{K}^0(-z)\rangle - |\bar{K}^0(z)K^0(-z)\rangle \right\}$$
$$= \frac{1}{\sqrt{2}} \left\{ |K_2^0(z)K_1^0(-z)\rangle - |K_1^0(z)K_2^0(-z)\rangle \right\}$$

for the two-kaon state before any decays have taken place. Since the initial state has $C = -1$, charge conjugation invariance in the strong decay of the Φ meson requires the negative sign between the two terms in the above expression. This description clearly includes quantum-mechanical interference between states which are spatially separated (see Sect. 7.1). Note that the above state vector is formally analogous to that of a spin-singlet state composed of two spin-1/2 particles: $|S = 0\rangle = [(\uparrow\downarrow) - (\downarrow\uparrow)]/\sqrt{2}$.

1.5 Strangeness Oscillations

"Especially interesting is the fact that we have taken the principle of superposition to its ultimately logical conclusion."

Richard Feynman, The Theory of Fundamental Processes

Suppose that at time $t = 0$ a pure K^0 beam is generated and then allowed to propagate in vacuum. After a sufficiently long period of time only the K_2^0 component will survive. This component is a superposition of the strangeness eigenstates K^0 and $\bar{K}^0$:

$$|K^0\rangle \longrightarrow |K_2^0\rangle = \frac{1}{\sqrt{2}} \left[|K^0\rangle - |\bar{K}^0\rangle \right], \quad t \gg \tau_1. \tag{1.13}$$

Thus starting out as a pure K^0 beam, the neutral kaon will evolve into a state of mixed strangeness. This is known as the K^0-$\bar{K}^0$, or strangeness, oscillation.

The first evidence for this effect was reported by K. Lande et al. in 1957 [5], who employed the same experimental set-up as was used to discover K_2^0 (see Fig. 1.2). They observed the process

$$\bar{K}^0 + \mathcal{H}e \rightarrow \Sigma^- p p n \pi^+$$

by using a beam of neutral kaons produced predominantly through reactions

$$p + n \rightarrow p + \Lambda^0 + K^0.$$

Since the energy threshold for the above interaction is much lower than for

$$p + n \rightarrow p + n + K^0 + \bar{K}^0$$

the beam overwhelmingly contained K^0 mesons.

The strangeness oscillation is a purely quantum-mechanical effect that enables one to test the principle of superposition of amplitudes in the most direct way. To learn more about this effect we first have to determine how the states of neutral kaons change in time.

The CP eigenstates K_1^0 and K_2^0 have different lifetimes and decay modes, and hence different weak couplings. Consequently, their masses also ought to differ, just as the mass difference betwen proton and neutron can be attributed to their different electromagnetic couplings.

A quantum-mechanical state (a particle) is described by a Schrödinger wave function, which is in general a complex number. The interference between two quantum states is determined by the difference in the phases of their wave functions. As we will see shortly, a K_1^0-K_2^0 mass difference causes the relative phase of K_1^0 and K_2^0 states to vary in time. It is precisely this time variation of the relative phase of the two-state system (K_1^0, K_2^0) that we want to study.

Suppose that at time $t = 0$ the state $|\Psi\rangle$ of a neutral kaon is pure K_1^0. A stable particle propagating in vacuum can be described by a plane wave $(\hbar = c = 1)$:

$$|\Psi(z,t)\rangle = e^{i(kz - Et)}|\Psi(0)\rangle, \tag{1.14}$$

where E is the energy of the particle and k its wave number. Since K_1^0 is not stable, we expect that at a later time the probability of finding the particle in the state $|\Psi(t = 0)\rangle$ should decrease by a factor e^{-t/τ_1} because of the exponential decay law for $K_1^0 \to \pi\pi$. This probability is given by

$$\text{Probability}(t) = |\text{Amplitude}(t)|^2 = \left|\langle K_1^0 \mid \Psi(t)\rangle\right|^2 = e^{-t/\tau_1}. \tag{1.15}$$

Therefore,

$$\text{Amplitude}(t) = \langle K_1^0 \mid \Psi(t)\rangle \propto e^{-t/2\tau_1} \tag{1.16}$$

and

$$|\Psi(t)\rangle = e^{-iE_1 t}\, e^{-t/2\tau_1}|K_1^0(t = 0)\rangle. \tag{1.17}$$

There is a similar expression for K_2^0. If t is measured in the paricle rest frame, then $E = m$, $\tau \equiv 1/\Gamma$ is the *proper lifetime* and

$$|\Psi_{1,2}(t)\rangle = e^{-(im_{1,2} + \Gamma_{1,2}/2)t_\text{p}}|K_{1,2}^0(t = 0)\rangle, \tag{1.18}$$

where t_p is the *proper time*.

An initially pure K^0-beam propagating in vacuum is thus described by

$$|\Psi(t)\rangle = \frac{1}{\sqrt{2}}\left[e^{-(im_1 + \Gamma_1/2)t_\text{p}}|K_1^0\rangle + e^{-(im_2 + \Gamma_2/2)t_\text{p}}|K_2^0\rangle\right] \tag{1.19}$$

The amplitude of the probability that the K-meson state $|\Psi(t)\rangle$ is a K^0 at time t reads

$$\mathsf{A} = \langle K^0 \mid \Psi(t)\rangle = \frac{1}{2}\left[e^{\phi_1} + e^{\phi_2}\right], \tag{1.20}$$

where

$$\phi = -(im + \Gamma/2)t_\text{p} \tag{1.21}$$

and Γ is the *decay rate*. Similarly, the probability amplitude for $\bar{K}^0$ is

$$\overline{\mathsf{A}} = \langle \bar{K}^0 \mid \Psi(t)\rangle = \frac{1}{2}\left[e^{\phi_1} - e^{\phi_2}\right] \tag{1.22}$$

The corresponding probabilities are

$$\mathsf{P} = \frac{1}{4}\left[e^{-\Gamma_1 t} + e^{-\Gamma_2 t} + 2e^{-(\Gamma_1 + \Gamma_2)t/2}\cos(\Delta m_k t)\right] \tag{1.23}$$

and

$$\overline{\mathsf{P}} = \frac{1}{4}\left[e^{-\Gamma_1 t} + e^{-\Gamma_2 t} - 2e^{-(\Gamma_1 + \Gamma_2)t/2}\cos(\Delta m_k t)\right], \tag{1.24}$$

where

$$\Delta m_k = |m_1 - m_2| \tag{1.25}$$

is the K_1^0-K_2^0 mass difference.

The K^0 and $\bar{K}^0$ intensities oscillate with the frequency Δm_k, as shown in Fig. 1.3. When a K^0 meson is created, the probability that it is a $\bar{K}^0$ is zero: $\mathsf{P}(t_0) = 1$ and $\overline{\mathsf{P}}(t_0) = 0$. As the K_1^0 component decays away, the original K^0 evolves into a state of mixed strangeness (see (1.13)). Each curve in Fig. 1.3 is the result of strangeness oscillation, associated with the cosine-term in (1.23), (1.24), superposed over exponential damping due to $\Gamma \neq 0$. The cosine term reflects quantum-mechanical interference between the K_1^0 and K_2^0 amplitudes of the particle over macroscopic distances!

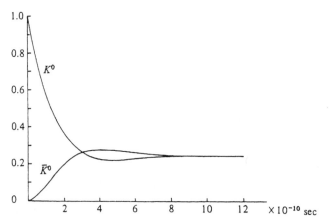

Fig. 1.3. Oscillations of K^0 and $\bar{K}^0$ intensities for an initially pure K^0 beam assuming $\Delta m_k = \Gamma_1/2$

Pioneering work on strangeness oscillations and the K_1^0-K_2^0 mass difference [6] reported a result in qualitative agreement with $\Delta m_k = \hbar/c^2\tau_1$. This experiment produced K^0 mesons by a low-energy π^- beam, and then detected $\bar{K}^0$ particles via hyperon production in a cloud chamber.

Again there is a nice optical analogy to this phenomenon: if we were to shine right-circularly polarized light through a crystal that absorbs predominantly x-polarized light, after a short distance there would be a large left-circularly polarized component. Both phenomena are based on the principle of superposition of amplitudes, which in quantum mechanics holds for "probability waves" of particles; these (complex) probability amplitudes can interfere constructively or destructively, just like waves do.

1.6 The K_1^0-K_2^0 Mass Difference

In the absence of the weak interaction, K^0 and $\bar{K}^0$ are two unconnected, degenerate ($m_{k^0} = m_{\bar{k}^0}$) states:

$$\langle \bar{K}^0 \mid \hat{H}_{\text{s+em}} \mid K^0 \rangle = \langle \bar{K}^0 \mid \hat{H}_{\text{s+em}}\hat{S} \mid K^0 \rangle = \langle \bar{K}^0 \mid \hat{S}\hat{H}_{\text{s+em}} \mid K^0 \rangle$$
$$= -\langle \bar{K}^0 \mid \hat{H}_{\text{s+em}} \mid K^0 \rangle,$$

i.e.

$$\langle \bar{K}^0 \mid \hat{H}_{\text{s+em}} \mid K^0 \rangle = 0 = \langle K^0 \mid \hat{H}_{\text{s+em}} \mid \bar{K}^0 \rangle. \tag{1.26}$$

In deriving the above result we used the fact that the hamiltonian $\hat{H}_{\text{s+em}}$ commutes with $\hat{S}$ (strangeness is conserved in strong and electromagnetic interactions).

In the particle rest frame, the masses of K^0 and $\bar{K}^0$ can be expressed as expectation values of $\hat{H}_{\text{s+em}}$ over the states $|K^0\rangle$ and $|\bar{K}^0\rangle$:

$$m_{k^0} = \langle K^0 \mid \hat{H}_{\text{s+em}} \mid K^0 \rangle = \langle \bar{K}^0 \mid \hat{H}_{\text{s+em}} \mid \bar{K}^0 \rangle = m_{\bar{k}^0} \tag{1.27}$$

The weak interaction, $\hat{H}_{\text{w}}$, is very feeble compared with $\hat{H}_{\text{s+em}}$, and can thus be treated as a small perturbation. As before, we will assume that $\hat{H}_{\text{w}}$ is strictly invariant under the combined operation of charge conjugation and parity transformation. The weak interaction connects K^0 and $\bar{K}^0$ (they are not eigenstates of the perturbed problem), but not the CP eigenstates K_1^0 and K_2^0:

$$\langle K_2^0 \mid \hat{H}_{\text{w}} \mid K_1^0 \rangle = 0 = \langle K_1^0 \mid \hat{H}_{\text{w}} \mid K_2^0 \rangle. \tag{1.28}$$

Note also that when a degeneracy exists, any linear combination of degenerate eigenstates is itself an eigenstate:

$$\hat{H}_{\text{s+em}}|\Psi\rangle = \hat{H}_{\text{s+em}} \left[a|K^0\rangle + b|\bar{K}^0\rangle\right] = m_{k^0}|\Psi\rangle,$$

where $|\Psi\rangle$ is the unperturbed state.

We can consequently apply ordinary *degenerate* perturbation theory to the K^0 system. Since neutral kaons decay through a number of channels, our Hilbert space should, in principle, be expanded to include all possible transitions. However, we keep the analysis simple by restricting ourselves to the two-dimensional Hilbert space spanned by $|K^0\rangle$ and $|\bar{K}^0\rangle$, in which case the effect of decays is incorporated into an *effective hamiltonian*

$$\hat{H}_{\text{w}}^{\text{eff}} = \hat{H}_{\text{w}} + \sum_n \frac{\hat{H}_{\text{w}}|n\rangle\langle n|\hat{H}_{\text{w}}}{m_{k^0} - E_n} + \cdots . \tag{1.29}$$

$\hat{H}_{\text{w}}^{\text{eff}}$ is determined by the virtual transitions to all intermediate states n outside the two-particle subspace. Now,

$$\left[\hat{H}_{\text{s+em}} + \hat{H}_{\text{w}}^{\text{eff}}\right]|\Psi\rangle = [m_{k^0} + m']|\Psi\rangle \tag{1.30}$$

where

$$a\hat{H}_{\mathrm{w}}^{\mathrm{eff}}|K^0\rangle + b\hat{H}_{\mathrm{w}}^{\mathrm{eff}}|\bar{K}^0\rangle = m'\left[a|K^0\rangle + b|\bar{K}^0\rangle\right] \tag{1.31}$$

is a small perturbation due to the weak interaction. Taking the inner product of (1.31) with $\langle K^0|$ and then $\langle \bar{K}^0|$, we obtain a system of two coupled linear equations:

$$\begin{aligned} a\langle K^0 \mid \hat{H}_{\mathrm{w}}^{\mathrm{eff}} \mid K^0\rangle + b\langle K^0 \mid \hat{H}_{\mathrm{w}}^{\mathrm{eff}} \mid \bar{K}^0\rangle &= am', \\ a\langle \bar{K}^0 \mid \hat{H}_{\mathrm{w}}^{\mathrm{eff}} \mid K^0\rangle + b\langle \bar{K}^0 \mid \hat{H}_{\mathrm{w}}^{\mathrm{eff}} \mid \bar{K}^0\rangle &= bm'. \end{aligned} \tag{1.32}$$

Since $\hat{H}_{\mathrm{w}}^{\mathrm{eff}}$ and $\hat{C}\hat{P}$ commute,

$$\begin{aligned} \langle \bar{K}^0 \mid \hat{H}_{\mathrm{w}}^{\mathrm{eff}} \mid K^0\rangle = \langle \bar{K}^0 \mid \hat{H}_{\mathrm{w}}^{\mathrm{eff}}\hat{C}\hat{P} \mid \bar{K}^0\rangle &= \langle \bar{K}^0 \mid \hat{C}\hat{P}\hat{H}_{\mathrm{w}}^{\mathrm{eff}} \mid \bar{K}^0\rangle \\ &= \langle K^0 \mid \hat{H}_{\mathrm{w}}^{\mathrm{eff}} \mid \bar{K}^0\rangle \equiv \Delta m \end{aligned} \tag{1.33}$$

and

$$\begin{aligned} \langle K^0 \mid \hat{H}_{\mathrm{w}}^{\mathrm{eff}} \mid K^0\rangle = \langle K^0 \mid \hat{H}_{\mathrm{w}}^{\mathrm{eff}}\hat{C}\hat{P} \mid \bar{K}^0\rangle &= \langle K^0 \mid \hat{C}\hat{P}\hat{H}_{\mathrm{w}}^{\mathrm{eff}} \mid \bar{K}^0\rangle \\ &= \langle \bar{K}^0 \mid \hat{H}_{\mathrm{w}}^{\mathrm{eff}} \mid \bar{K}^0\rangle \equiv m. \end{aligned} \tag{1.34}$$

Expressed in matrix form, (1.32) now reads

$$\begin{pmatrix} m - m' & \Delta m \\ \Delta m & m - m' \end{pmatrix} \begin{pmatrix} a \\ b \end{pmatrix} = 0. \tag{1.35}$$

To find nontrivial solutions of (1.35) we set the determinant of the coefficients of a and b equal to zero, with the result $m' = m \pm \Delta m$. Inserting each of the values of m' into (1.35) yields

$$\begin{aligned} a/b &= 1, & m' &= m + \Delta m, \\ a/b &= -1, & m' &= m - \Delta m. \end{aligned} \tag{1.36}$$

To ensure that the new $|\Psi\rangle$ is normalized ($|a|^2 + |b|^2 = 1$), we set $a = 1/\sqrt{2}$ and $b = \pm 1/\sqrt{2}$. Therefore,

$$|\Psi\rangle_{\mathrm{new}} = a|K^0\rangle + b|\bar{K}^0\rangle = \begin{pmatrix} |K_1^0\rangle \\ |K_2^0\rangle \end{pmatrix}, \tag{1.37}$$

where K_1^0 and K_2^0 are the correct linear superpositions of the original eigenfunctions which diagonalize the perturbation.

The "weak" masses of the two nondegenerate levels K_1^0 and K_2^0 are given by[8]

$$\begin{aligned} \langle K_1^0 \mid \hat{H}_{\mathrm{w}}^{\mathrm{eff}} \mid K_1^0\rangle &= \frac{1}{2}\left[\langle K^0| + \langle \bar{K}^0|\right] \hat{H}_{\mathrm{w}}^{\mathrm{eff}} \left[|K^0\rangle + |\bar{K}^0\rangle\right] \\ &= m + \Delta m \end{aligned} \tag{1.38}$$

---

[8] Note that $m_1 = m - \Delta m$ and $m_2 = m + \Delta m$ if we choose $\theta_{\mathrm{cp}} = \pi$ (see footnote 5). Experimentally, $m_2 > m_1$, as shown in Sect. 2.3.

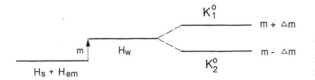

Fig. 1.4. The K^0-$\bar{K}^0$ mixing results in level splitting

and

$$\langle K_2^0 \mid \hat{H}_{\mathrm{w}}^{\mathrm{eff}} \mid K_2^0 \rangle = \frac{1}{2}\left[\langle K^0| - \langle \bar{K}^0|\right] \hat{H}_{\mathrm{w}}^{\mathrm{eff}} \left[|K^0\rangle - |\bar{K}^0\rangle\right]$$
$$= m - \Delta m \tag{1.39}$$

The above perturbation calculation can be summarized as shown in Fig. 1.4. The weak interaction shifts m_{k^0} by m and splits the degenerate levels by an amount $2\Delta m \equiv \Delta m_k$. The two new levels with definite CP parity, K_1^0 and K_2^0, differ in mass:

$$\Delta m_k \equiv |m_1 - m_2| = \langle K^0 \mid \hat{H}_{\mathrm{w}}^{\mathrm{eff}} \mid \bar{K}^0 \rangle + \langle \bar{K}^0 \mid \hat{H}_{\mathrm{w}}^{\mathrm{eff}} \mid K^0 \rangle. \tag{1.40}$$

This mass difference is due to strangeness-changing K^0-$\bar{K}^0$ transitions. We can make a rather accurate estimate of Δm_k based on the Heisenberg uncertainty relation, $\Delta E\,\Delta t = 1$ ($\hbar = c = 1$), by setting $\Delta t = 1/(\Gamma_1 + \Gamma_2) \approx 1/\Gamma_1$ and $\Delta E = \Delta m_k$ (cf. Sect. 9.1):

$$\Delta m_k \approx \Gamma_1 \approx 10^{10}\ \mathrm{s}^{-1} \approx 7 \times 10^{-6}\ \mathrm{eV}. \tag{1.41}$$

The K_1^0-K_2^0 mass difference can be obtained by counting the number of $\bar{K}^0$-induced events as a function of their distance from the K^0 production point (see (1.24)). In the experiment [7], for example, a propane bubble chamber was used to detect the following sequence of events:

$$\overbrace{K^+ + n \rightarrow p + K^0,}^{\text{charge-exchange}}\ \ K^0 \rightarrow K_2^0 = K^0 + \bar{K}^0,$$
$$\bar{K}^0 + p \rightarrow \Lambda^0 + \pi^+. \tag{1.42}$$

The measured value of Δm_k is indeed tiny:

$$\Delta m_k = 3.5 \times 10^{-6}\ \mathrm{eV} \approx \Gamma_1/2 \quad \text{or} \quad \frac{\Delta m_k}{m_k} = 0.7 \times 10^{-14}. \tag{1.43}$$

The smallness of the observed K_1^0-K_2^0 mass difference indicates that $\langle K^0 \mid \hat{H}_{\mathrm{w}} \mid \bar{K}^0 \rangle = \langle \bar{K}^0 \mid \hat{H}_{\mathrm{w}} \mid K^0 \rangle = 0$, thus confirming that first-order weak interactions obey the empirical rule $|\Delta S| \le 1$.

A much more precise way of measuring Δm_k is based on the K_1^0 *regeneration phenomenon*, another spectacular quantum-mechanical effect associated with the K^0 system that we describe next in some detail.

2. Propagation of Neutral Kaons in Matter

2.1 The K_1^0 Regeneration

*"I believe that this concept of probability amplitude is perhaps the most funda-
mental concept of quantum theory."*

P.A.M. Dirac, Relativity and Quantum Theory

This phenomenon — the conversion of K_2^0 mesons into K_1^0 mesons — is of the
same nature as the K^0-$\bar{K}^0$ oscillation. Its existence was predicted by A. Pais
and O. Piccioni in 1955 [8], a year before the K_2^0 particle was discovered! A
detailed analysis of the process was presented in the seminal paper by M. L.
Good [8].

Suppose that a pure K^0 beam is allowed to decay in vacuum until we are
left with only the K_2^0 component. If we then pass this K_2^0 beam through a
thin slab of material where it can interact, the strong interactions will select
both eigenstates of strangeness present in the beam. The $\bar{K}^0$ component will
be more strongly absorbed than the K^0 component because K^0 mesons can
only scatter elastically or through charge-exchange reactions, whereas $\bar{K}^0$
mesons can be both scattered and absorbed (see (1.2),(1.3)). After the slab,
the beam composition can be expressed as

$$
\begin{aligned}
|\Psi_{\text{after}}\rangle &= \frac{1}{\sqrt{2}}\left[a|K^0\rangle - b|\bar{K}^0\rangle\right] \\
&= \frac{a+b}{2\sqrt{2}}\left[|K^0\rangle - |\bar{K}^0\rangle\right] + \frac{a-b}{2\sqrt{2}}\left[|K^0\rangle + |\bar{K}^0\rangle\right] \\
&= \frac{1}{2}(a+b)|K_2^0\rangle + \frac{1}{2}(a-b)|K_1^0\rangle.
\end{aligned}
\tag{2.1}
$$

Since $a \neq b$, we conclude that after all the K_1^0 mesons in the initial beam have
decayed away, some can be *regenerated* by passing the pure K_2^0 beam through
matter. Figure 2.1 contains a schematic drawing of such an experiment.

Like the K^0-$\bar{K}^0$ oscillation, the K_1^0 regeneration phenomenon is a direct
consequence of the principle of superposition of amplitudes in quantum mechanics. The analogous experiment with polarized light is shown in Fig. 2.2.
This analogy follows from the wave aspect of quantum theory.

There is another close analogy to the K_1^0 regeneration which beautifully illustrates the concept of quantization. It is based on the Stern–Gerlach atomic

Fig. 2.1. Schematic drawing of a K_1^0 regeneration experiment

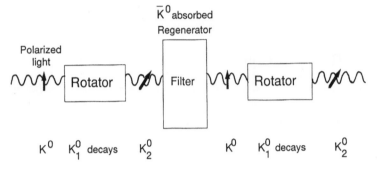

Fig. 2.2. An experiment with polarized light analogous to K_1^0 regeneration

beam experiment, i.e., on the fact that it is impossible to quantize the spin components of the beam along two orthogonal axis simultaneously.

As shown in Fig. 2.3, an unpolarized atomic beam of spin 1/2 propagating in the z direction enters an inhomogeneous magnetic field that points along the y axis (H_y). This causes the beam to split equally into two components, one deflected upward and the other downward. The two components correspond to the atoms quantized by the field in the spin eigenstates $\sigma_y = +1/2$ and $\sigma_y = -1/2$, respectively. If the "lower" component is then sent through an inhomogeneous magnetic field pointing in the direction perpendicular to the y axis (the x direction), it will again split into two equal components, one deflected in the x direction ($\sigma_x = +1/2$) and the other in the opposite

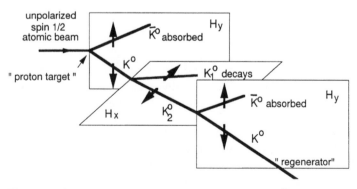

Fig. 2.3. Analogy between the Stern–Gerlach and K_1^0 regeneration experiments

direction ($\sigma_x = -1/2$). We now repeat the above two steps starting with the $\sigma_x = -1/2$ component, and assume that after each beam-splitting the $\sigma_{x,y} = +1/2$ component is absorbed in matter.

The analogy between this and the K_1^0 regeneration experiment is as follows. Each time the atomic beam passes through H_y corresponds to the selection of strangeness eigenstates K^0 and $\bar{K}^0$ via strong interactions in the proton target or the *regenerator*. The beam splitting in H_x is analogous to the selection of CP eigenstates K_1^0 and K_2^0 through weak decays.

Note that once the beam passes through an inhomogeneous magnetic field in, say, the y direction, all preexisting information about quantization along the x axis is lost (the axis singled out in space is defined by the magnetic field). This means that it is impossible to quantize the spin components of the beam along two orthogonal axes; in other words, the spin operators $\hat{\sigma}_x$ and $\hat{\sigma}_y$ do not commute. Similarly, the operators $\hat{S}$ and $\hat{C}\hat{P}$ also do not commute; hence K^0 states are eigenstates of either $\hat{S}$ or $\hat{C}\hat{P}$, but not both.

The first experimental confirmation of the regeneration phenomenon, and also the first measurement of the K_1^0-K_2^0 mass difference, was that by O. Piccioni and his collaborators [9]. To produce K_2^0 mesons, they passed negative pions through a liquid-hydrogen target (see (1.1)), and then allowed the K_1^0 component of the K^0 beam to decay away. About 200 K_1^0 mesons were regenerated in a 30-inch propane bubble chamber fitted with lead and iron plates (see Fig. 2.4), from a beam of approximately 10^5 K_2^0 particles.

A particularly elegant way of measuring Δm_k is based on the *variable gap method*. Refer to Fig. 2.5. A K_2^0 beam passes through two thin slabs of material separated by distance g, which can be varied. Let us assume for the sake of simplicity that the $\bar{K}^0$ component is totally absorbed in each of the two slabs. After exiting from the first slab ($t = 0$), the beam is pure K^0.

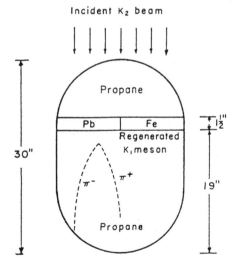

Incident K_2 beam

Propane

Pb   Fe

Regenerated K_1 meson

π^- π^+

Propane

30"

1½"

19"

Fig. 2.4. A regenerated K_1^0 decays in the experiment by R. Good et al. [9]

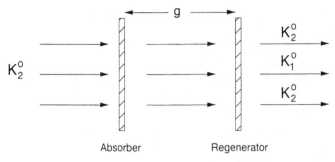

Fig. 2.5. A sketch of the variable gap method

Just before it enters the second slab (*regenerator*), the beam is described by
(1.19), where $t_p \equiv t_g = t/\gamma = g/\beta\gamma = g/[(p_k/E_k)(E_k/m_k)] = g(m_k/p_k)$ is
the proper time to traverse g. Immediately after the regenerator ($t_g \gg t_{\text{slab}}$),
where the $\bar{K}^0$ component is totally absorbed, the K_1^0 amplitude reads (see
(1.20))

$$\mathsf{A}_{K_1^0}(t_g) = \langle K_1^0 \mid \Psi(t_g)\rangle = \frac{1}{\sqrt{2}}\langle K^0 \mid \Psi(t_g)\rangle = \frac{1}{2\sqrt{2}}\left[e^{\phi_1} + e^{\phi_2}\right] \qquad (2.2)$$

with $\phi_{1,2}$ given by (1.21). The corresponding decay intensity is ($\Gamma_1 \gg \Gamma_2$)

$$\mathsf{I}(g) \equiv \left|\mathsf{A}_{K_1^0}\right|^2 \approx \frac{\mathsf{I}(0)}{4}\left[1 + e^{-\Gamma_1 t_g} + 2e^{-\Gamma_1 t_g/2}\cos(\Delta m_k t_g)\right], \qquad (2.3)$$

where

$$\mathsf{I}(0) = \left|\langle K_1^0 \mid \Psi(0)\rangle\right|^2 = \frac{1}{2}. \qquad (2.4)$$

The mass difference Δm_k can be obtained by measuring the $K_1^0 \to \pi^+\pi^-$
decay intensity after the regenerator as a function of g. This is, of course,
a very simplified description of the method (see [10] regarding the original
variable gap experiment).

2.2 Coherent Regeneration Amplitude

We have seen that regeneration occurs because K^0 and $\bar{K}^0$ mesons interact
differently with matter. Let us rewrite (2.1) as

$$|\Psi_{\text{after}}\rangle = \frac{a+b}{2}\left[|K_2^0\rangle + \varrho|K_1^0\rangle\right], \quad \varrho \equiv \frac{a-b}{a+b}, \qquad (2.5)$$

where ϱ, the *regeneration parameter*, is a complex number that can be related
to the physical properties of the regenerator, as will be shown in what follows.

The regenerated K_1^0 mesons will coincide in direction and momentum
with the incident K_2^0 beam whenever the *forward scattering* amplitudes for

K^0 and $\bar{K}^0$ are different. This situation is called *coherent regeneration*, or *regeneration by transmission*, and is a result of the constructive contribution from all scatterers in the target. The coherence is preserved in a region that is typically contained within a micro-radian. As in the case of forward scattering of neutrons, light, X-rays, etc., the forward scattering of neutral kaons is always coherent, and its interference with the incoming beam gives rise to the *refractive index* of the scattering material (see Appendix B):

$$n = 1 + \frac{2\pi N}{k^2} \mathfrak{f}(\omega, 0), \tag{2.6}$$

where N is the number of scatterers per unit volume, k the wave number of the incident particles and $\mathfrak{f}(\omega, 0)$ the scattering amplitude in the forward direction ($\theta = 0$). The forward amplitude is related to the total cross-section by the *optical theorem*

$$\sigma_{\text{tot}} = \frac{4\pi}{k} \operatorname{Im} \mathfrak{f}(\omega, 0). \tag{2.7}$$

A particle traversing a slab of thickness z picks up an extra phase, which is proportional to the refractive index n:

$$\varphi = k(n-1)z. \tag{2.8}$$

Not only are the K^0 and $\bar{K}^0$ mesons absorbed differently in matter, but their elastic scattering amplitudes also differ, just as those for $K^+p \to K^+p$ and $K^-p \to K^-p$ do. Since K^0 and $\bar{K}^0$ have different total cross-sections, they must also have different indices of refraction; hence they acquire unequal phase shifts while propagating through matter. To see what this implies, we first write down the time development of K^0 and $\bar{K}^0$ states in vacuum (see (1.19)):

$$|K^0(t_{\text{p}})\rangle = \frac{1}{\sqrt{2}} \left[|K_1^0\rangle \, \mathrm{e}^{\phi_1} + |K_2^0\rangle \, \mathrm{e}^{\phi_2} \right],$$

$$|\bar{K}^0(t_{\text{p}})\rangle = \frac{1}{\sqrt{2}} \left[|K_1^0\rangle \, \mathrm{e}^{\phi_1} - |K_2^0\rangle \, \mathrm{e}^{\phi_2} \right], \tag{2.9}$$

where

$$\phi_{1,2} = -\left(\mathrm{i}m_{1,2} + \Gamma_{1,2}/2\right) t_{\text{p}} \tag{2.10}$$

and $t_{\text{p}} = t/\gamma = t\sqrt{1-v^2}$ is the proper time: $z = vt = t_{\text{p}}v/\sqrt{1-v^2}$. The two-state Schrödinger equation describing the system (2.9) reads[9]

$$\mathrm{i}\frac{\mathrm{d}\Psi_{\text{vac}}}{\mathrm{d}t_{\text{p}}} = \begin{pmatrix} \tilde{m} - \mathrm{i}\tilde{\Gamma}/2 & \delta m - \mathrm{i}\delta\Gamma/2 \\ \delta m - \mathrm{i}\delta\Gamma/2 & \tilde{m} - \mathrm{i}\tilde{\Gamma}/2 \end{pmatrix} \Psi_{\text{vac}}, \tag{2.11}$$

---

[9] The K^0 system is described by two coupled differential equations because of the $K^0 \leftrightarrow \bar{K}^0$ mixing. If neutral kaons interacted only through strong and electromagnetic interactions, both of which conserve S, there would be no transitions between the K^0 and $\bar{K}^0$ states, and each state would be separately described by a Schrödinger equation. In fact, (2.11) is not a "real" Schrödinger equation.

where

$$\Psi_{\text{vac}} = \begin{pmatrix} |K^0(t_{\text{p}})\rangle \\ |\bar{K}^0(t_{\text{p}})\rangle \end{pmatrix} \qquad (2.12)$$

and

$$\tilde{m} = \frac{m_1 + m_2}{2}, \quad \tilde{\Gamma} = \frac{\Gamma_1 + \Gamma_2}{2}, \quad \delta m = \frac{m_1 - m_2}{2}, \quad \delta\Gamma = \frac{\Gamma_1 - \Gamma_2}{2}. \qquad (2.13)$$

While traversing a distance z in matter, the $K^0(\bar{K}^0)$ meson acquires an extra phase:

$$|K^0(t_{\text{p}})\rangle \to |K^0(t_{\text{p}})\rangle \, e^{i\varphi}, \quad |\bar{K}^0(t_{\text{p}})\rangle \to |\bar{K}^0(t_{\text{p}})\rangle \, e^{i\bar{\varphi}}, \qquad (2.14)$$

where

$$\varphi = k(n-1)z = \frac{2\pi N v}{k\sqrt{1-v^2}} t_{\text{p}} f(0),$$

$$\bar{\varphi} = k(\bar{n}-1)z = \frac{2\pi N v}{k\sqrt{1-v^2}} t_{\text{p}} \bar{f}(0). \qquad (2.15)$$

It is straightforward to show that (2.11) now becomes

$$i\frac{d\Psi_{\text{matt}}}{dt_{\text{p}}}$$

$$= \begin{pmatrix} \tilde{m} - i\tilde{\Gamma}/2 - \frac{2\pi N v}{k\sqrt{1-v^2}} f(0) & \delta m - i\delta\Gamma/2 \\ \delta m - i\delta\Gamma/2 & \tilde{m} - i\tilde{\Gamma}/2 - \frac{2\pi N v}{k\sqrt{1-v^2}} \bar{f}(0) \end{pmatrix} \Psi_{\text{matt}}. \qquad (2.16)$$

The matrix equation (2.16) may be expressed as

$$i\frac{d\Psi_{\text{matt}}}{dt_{\text{p}}} = (M - i\Gamma)\Psi_{\text{matt}} - \frac{2\pi N v}{k\sqrt{1-v^2}} \begin{pmatrix} f(0) & 0 \\ 0 & \bar{f}(0) \end{pmatrix} \Psi_{\text{matt}} \qquad (2.17)$$

or

$$i\frac{d\Psi_{\text{matt}}}{dt_{\text{p}}} = \left(M' - i\Gamma' \right) \Psi_{\text{matt}}, \qquad (2.18)$$

with

$$M' - i\Gamma' \equiv M - i\Gamma - \xi \begin{pmatrix} f(0) & 0 \\ 0 & \bar{f}(0) \end{pmatrix} \qquad (2.19)$$

and

$$\xi \equiv \frac{2\pi N v}{k\sqrt{1-v^2}}. \qquad (2.20)$$

The time development of the K^0 system in vacuum and in matter is therefore described by the Schrödinger equations

$$i\frac{d\Psi_{\text{vac}}}{dt_{\text{p}}} = H\Psi_{\text{vac}} \quad \text{and} \quad i\frac{d\Psi_{\text{matt}}}{dt_{\text{p}}} = H'\Psi_{\text{matt}}, \qquad (2.21)$$

respectively, with

$$H \equiv M - i\Gamma, \quad H^{'} \equiv M^{'} - i\Gamma^{'}, \tag{2.22}$$

where M and Γ are 2×2 matrices.[10] M is called the *mass matrix* and Γ the *decay matrix*; both are hermitian ($M_{jk}^* = M_{kj}$, $\Gamma_{jk}^* = \Gamma_{kj}$) because they represent observable quantities. However, H and $H^{'}$ are not hermitian, otherwise K^0 and $\bar{K}^0$ would not decay. To understand this statement note that the decay process is the equivalent of an absorption, and that an absorbing medium can be described in terms of a complex index of refraction, the imaginary part of which is associated with attenuation (see Appendix B).

We may now proceed with our main task, which is to obtain the intensity of the regenerated K_1^0 component at z. To this end we express Ψ_{matt} in (2.16) in terms of the CP eigenstates K_1^0 and K_2^0 and write the result down as a system of two coupled differential equations ($|K_{1,2}^0\rangle \equiv K_{1,2}^0$):

$$\begin{aligned}
i\frac{d}{dt_p}\left[K_1^0 + K_2^0\right] &= \left(\tilde{m} - i\tilde{\Gamma}/2 - \xi\mathfrak{f}\right)\left[K_1^0 + K_2^0\right] \\
&\quad + (\delta m - i\delta\Gamma/2)\left[K_1^0 - K_2^0\right], \\
i\frac{d}{dt_p}\left[K_1^0 - K_2^0\right] &= (\delta m - i\delta\Gamma/2)\left[K_1^0 + K_2^0\right] \\
&\quad + \left(\tilde{m} - i\tilde{\Gamma}/2 - \xi\bar{\mathfrak{f}}\right)\left[K_1^0 - K_2^0\right].
\end{aligned} \tag{2.23}$$

Adding and subtracting equations (2.23) yields

$$i\frac{d\Psi_{1,2}}{dt_p} = \begin{pmatrix} m_1 - i\Gamma_1/2 & 0 \\ 0 & m_2 - i\Gamma_2/2 \end{pmatrix}\Psi_{1,2} - \frac{\xi}{2}\begin{pmatrix} \mathfrak{f} + \bar{\mathfrak{f}} & \mathfrak{f} - \bar{\mathfrak{f}} \\ \mathfrak{f} - \bar{\mathfrak{f}} & \mathfrak{f} + \bar{\mathfrak{f}} \end{pmatrix}\Psi_{1,2} \tag{2.24}$$

with

$$\Psi_{1,2} \equiv \begin{pmatrix} K_1^0 \\ K_2^0 \end{pmatrix}. \tag{2.25}$$

We see that $d\Psi_{1,2}/dt_p$ has two components: the first one describes the propagation of free particles, and the second one the scattering with nuclei inside the regenerator. The change of $\Psi_{1,2}$ with respect to z as the beam passes through matter may thus be expressed as

$$\frac{d\Psi_{1,2}}{dz} = \left[\frac{d\Psi_{1,2}}{dz}\right]_{\mathrm{vac}} + \left[\frac{d\Psi_{1,2}}{dz}\right]_{\mathrm{nucl}}. \tag{2.26}$$

The second term in (2.26) is readily obtained from (2.24) by setting $dz = vdt = v\gamma dt_p = vdt_p/\sqrt{1 - v^2}$:

---

[10] Since we have restricted ourselves to a two-dimensional Hilbert space, H and $H^{'}$ are *effective hamiltonians*: $H = H_{\mathrm{strong}} + H_{\mathrm{electromagnetic}} + H_{\mathrm{weak}}$. From (2.22) it follows that $M = (H + H^\dagger)/2$ and $\Gamma = i(H - H^\dagger)$.

$$\left[\frac{d\Psi_{1,2}}{dz}\right]_{\text{nucl}} = \frac{i\pi N}{k}\begin{pmatrix} f+\bar{f} & f-\bar{f} \\ f-\bar{f} & f+\bar{f} \end{pmatrix}\Psi_{1,2}. \tag{2.27}$$

Regarding the first term, recall that an unstable particle propagating in vacuum may be described by a plane wave

$$\Psi(z) = e^{ikz - z/2\Lambda}\Psi(0), \tag{2.28}$$

where we used $\Gamma t_{\mathrm{p}}/2 = z/2\beta\gamma\tau = z/2\Lambda$ ($\Lambda \equiv \beta\gamma\tau$ is the decay length). Therefore,

$$\left[\frac{d\Psi_{1,2}}{dz}\right]_{\text{vac}} = \begin{pmatrix} ik_1 - 1/2\Lambda_1 & 0 \\ 0 & ik_2 - 1/2\Lambda_2 \end{pmatrix}\Psi_{1,2}. \tag{2.29}$$

Expressions (2.27) and (2.29) form a system of two coupled differential equations:

$$\begin{aligned} \frac{dK_1^0}{dz} &= \left[ik_1 - 1/2\Lambda_1 + \frac{i\pi N}{k_1}\left(f+\bar{f}\right)\right]K_1^0 + \frac{i\pi N}{k_1}\left(f-\bar{f}\right)K_2^0 \\ &\equiv \mathcal{C}K_1^0 + \mathcal{D}K_2^0, \\ \frac{dK_2^0}{dz} &= \left[ik_2 - 1/2\Lambda_2 + \frac{i\pi N}{k_2}\left(f+\bar{f}\right)\right]K_2^0 + \frac{i\pi N}{k_2}\left(f-\bar{f}\right)K_1^0 \\ &\equiv \mathcal{A}K_2^0 + \mathcal{B}K_1^0. \end{aligned} \tag{2.30}$$

The coefficients $\mathcal{A}$, $\mathcal{B}$, $\mathcal{C}$ and $\mathcal{D}$ in (2.30) represent the rates of change of the K_1^0 and K_2^0 amplitudes due to propagation, decay and (coherent) scattering. The coupled differential equations (2.30) can be formally solved for the z dependence of the amplitudes. However, the solution is simplified, and underlying physics made more transparent, by noting that $K_2^0 \gg K_1^0$ and $\Gamma_1 \gg \Gamma_2$. We can thus neglect the contribution of K_1^0 to the change in K_2^0 and write the second equation in (2.30) as

$$\frac{dK_2^0}{K_2^0} \approx \left[ik_2 - 1/2\Lambda_2 + \frac{i\pi N}{k_2}\left(f+\bar{f}\right)\right]dz. \tag{2.31}$$

This yields

$$\ln K_2^0(z) \xrightarrow{\Lambda_2 \gg z} ik_2 z + \frac{i\pi N}{k_2}\left(f+\bar{f}\right)z + \ln K_2^0(0), \tag{2.32}$$

i.e.

$$K_2^0(z) = e^{ik_2 z}\, e^{-Nf_{\text{tot}}z/2}\, K_2^0(0), \tag{2.33}$$

where we defined[11]

---

[11] Writing $\sigma_{\text{tot}}(K_1^0) = \sigma_{\text{tot}}(K_2^0) = [\sigma_{\text{tot}}(K^0) + \sigma_{\text{tot}}(\bar{K}^0)]/2 = 4\pi[\text{Im}\,f + \text{Im}\,\bar{f}]/2k = -4\pi i[i\text{Im}\,f + i\text{Im}\,\bar{f}]/2k$ we could identify f_{tot} with σ_{tot} provided $i\text{Im}\,f = f$ and $i\text{Im}\,\bar{f} = \bar{f}$, i.e. the forward scattering amplitudes were purely imaginary (the beam was only attenuated). In general this is not the case because of refraction.

$$f_{\text{tot}} \equiv -\frac{4\pi \mathrm{i}}{k} \frac{\mathfrak{f} + \bar{\mathfrak{f}}}{2}. \tag{2.34}$$

Inserting (2.33) into the first equation (2.30) gives

$$\frac{\mathrm{d}K_1^0(z)}{\mathrm{d}z} = \left[\mathrm{i}k_1 - 1/2\varLambda_1 + \frac{\mathrm{i}\pi N}{k_1}\left(\mathfrak{f} + \bar{\mathfrak{f}}\right)\right] K_1^0(z)$$

$$+ \frac{\mathrm{i}\pi N}{k_1}\left(\mathfrak{f} - \bar{\mathfrak{f}}\right) \mathrm{e}^{\mathrm{i}k_2 z}\, \mathrm{e}^{-N f_{\text{tot}} z/2}\, K_2^0(0) \tag{2.35}$$

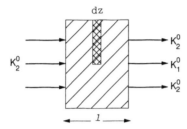

Fig. 2.6. Schematic drawing of a solid regenerator of thickness l

A K_1^0 travels as a K_2^0 with a wave number k_2 until it is regenerated, and after that as a K_1^0 with a wave number k_1. Since Δm_k is so tiny, $k_1 \approx k_2$. Note also that $\varLambda_{\text{vac}} \approx \varLambda_{\text{matt}}$. The first term on the right-hand side of (2.35) describes the behavior of K_1^0 mesons, created before z, in the interval $z, z+\mathrm{d}z$. The second term gives the amplitude of K_1^0 particles created between z and $z + \mathrm{d}z$ (see Fig. 2.6). To see how this amplitude is modified at the end of the regenerator due to propagation, decay and coherent scattering, we integrate the first term in the interval $l - z$, where l is the thickness of the regenerator, with the result

$$\mathcal{C}(l - z) = \mathrm{e}^{\mathrm{i}k_1(l-z)}\, \mathrm{e}^{-(l-z)/2\varLambda_1}\, \mathrm{e}^{-N f_{\text{tot}}(l-z)/2}. \tag{2.36}$$

Since we are dealing with coherent scattering, the amplitudes of K_1^0 mesons created in all $\mathrm{d}z$ intervals have to be added. This gives the following K_1^0 amplitude at the end of the regenerator:

$$K_1^0(l) = \int_0^l \mathcal{C}(l-z)\mathrm{d}\mathcal{D}(z)$$

$$= \frac{\mathrm{i}\pi N}{k_1}\left(\mathfrak{f} - \bar{\mathfrak{f}}\right) K_2^0(0)$$

$$\times \int_0^l \mathrm{e}^{\mathrm{i}k_1(l-z)}\, \mathrm{e}^{-(l-z)/2\varLambda_1}\, \mathrm{e}^{-N f_{\text{tot}}(l-z)/2}\, \mathrm{e}^{\mathrm{i}k_2 z}\, \mathrm{e}^{-N f_{\text{tot}} z/2}\, \mathrm{d}z$$

$$= \frac{\mathrm{i}\pi N}{k_1}\left(\mathfrak{f} - \bar{\mathfrak{f}}\right) \mathrm{e}^{-N f_{\text{tot}} l/2}\, \mathrm{e}^{\mathrm{i}k_2 l}\, K_2^0(0)$$

$$\times \int_0^l \mathrm{e}^{\mathrm{i}(k_1-k_2)(l-z)}\, \mathrm{e}^{-(l-z)/2\varLambda_1}\, \mathrm{d}z$$

$$= \frac{\mathrm{i}\pi N}{k_1}\left(\mathfrak{f} - \bar{\mathfrak{f}}\right) K_2^0(l) \int_0^l \mathrm{e}^{(l-z)[\mathrm{i}(k_1-k_2)-1/2\varLambda_1]}\, \mathrm{d}z \tag{2.37}$$

with

$$K_2^0(l) \equiv e^{-Nf_{tot}l/2} e^{ik_2l} K_2^0(l=0). \tag{2.38}$$

The simple integration above results in

$$K_1^0(l) = \frac{iN\lambda_1\Lambda_1\mathfrak{f}_r}{-i\delta_m + 1/2} \left[1 - e^{-(-i\delta_m + 1/2)\zeta}\right] K_2^0(l) \equiv \varrho_c K_2^0(l), \tag{2.39}$$

where

$$\mathfrak{f}_r \equiv (\mathfrak{f} - \bar{\mathfrak{f}})/2, \quad \lambda_1 \equiv 2\pi/k_1, \quad \zeta \equiv l/\Lambda_1, \tag{2.40}$$

and

$$\delta_m \equiv \frac{(m_2 - m_1)c^2}{\hbar/\tau_1} = (k_1 - k_2)\Lambda_1 \tag{2.41}$$

is the K_2^0-K_1^0 mass difference expressed in units of the K_1^0 mass width $\hbar/\tau_1$. Expression (2.41) requires a few words of explanation. As discussed in Appendix B, for coherent scattering the regenerator as a whole absorbs the difference in momentum $p_1 - p_2 = \hbar(k_1 - k_2) = \hbar\Delta k$ and recoils with the momentum $-\hbar\Delta k$, taking away from the meson the energy $(\hbar\Delta k)^2/2M_{reg}$. Since M_{reg} is very large, this energy is negligible compared with $m_1 - m_2$. Hence $E_2 = E_1$, which we express as $(\hbar, c \neq 1)$

$$\hbar^2\left(k_1^2 - k_2^2\right) = \left(m_2^2 - m_1^2\right)c^2$$

i.e.

$$m_2 - m_1 = \frac{k_1 - k_2}{mc^2} k\hbar^2,$$

where $k \equiv (k_1 + k_2)/2$ and $m \equiv (m_1 + m_2)/2$ can be taken as the momentum and mass of either kaon. Expression (2.41) follows from the above result if we note that k (like $\mathfrak{f}$) is determined in the rest-frame of the scatterers: $k = p/\hbar = \beta E/\hbar = \beta\gamma m/\hbar$.

Equation (2.39) defines the *coherent regeneration amplitude* ϱ_c. The intensity of K_1^0 mesons emerging from the regenerator is thus given by

$$\begin{aligned}
I_{K_1^0}(l) &= \frac{(N\lambda_1\Lambda_1)^2}{\delta_m^2 + 1/4} |\mathfrak{f}_r|^2 \left[1 + e^{-\zeta} - 2e^{-\zeta/2}\cos(\delta_m\zeta)\right] \left|K_2^0(0)\right|^2 e^{-Nf_{tot}l} \\
&= |\varrho_c|^2 I_{K_2^0}(l). \tag{2.42}
\end{aligned}$$

Note that expression (2.42) has the same oscillatory form as that for the $K^0(\bar{K}^0)$ intensity (see (1.23), (1.24)). This is to be expected since in both cases oscillations arise from the interference of the two eigenstates of the hamiltonian, which have the same masses and lifetimes in the two cases. The above results are summarized in Fig. 2.7.

The magnitude of δ_m can be determined by measuring ϱ_c as a function of regenerator thickness. This method is relatively simple and does not require a knowledge of $\mathfrak{f}_r$. Using an iron regenerator, a spark chamber experiment

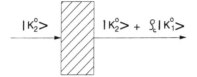

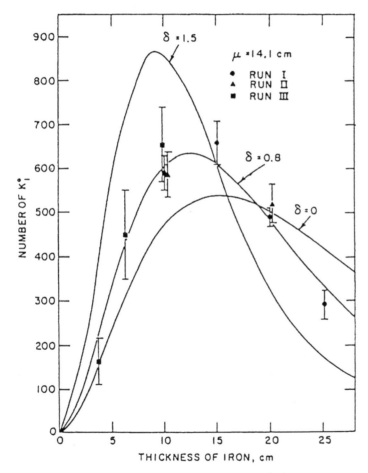

Fig. 2.7. K^0 beam composition before and immediately after a regenerator

[11] obtained $\delta_m = 0.82 \pm 0.12$ by plotting the K_1^0 intensity as a function of the iron thickness l (see Fig. 2.8). Since

$$I_{K_2^0}(l) \propto e^{-Nf_{\text{tot}}l} \equiv e^{-l/\mu} \tag{2.43}$$

they measured the nuclear mean path, μ, in a separate attenuation experiment.

Fig. 2.8. The magnitude of Δm_k measured by [11]

2.3 K_1^0-K_1^0 Interference and the Sign of Δm_k

We next describe a regeneration experiment which determined the magnitude, as well as the sign, of the K_1^0-K_2^0 mass difference Δm_k. The measurement was based on the observation of the interference between the regenerated (reg) and originally produced (orig) K_1^0 mesons, which proves unequivocally that the states K_1^0(reg) and K_1^0(orig) are quantum-mechanically identical [12].

Consider a pure K^0 beam impinging on a regenerator of thickness l placed at a distance d from the production point. Both l and d can be so adjusted that the two interfering waves, K_1^0(orig) and K_1^0(reg), have comparable amplitudes, thus maximizing the interference effect. At the exit of the regenerator

$$\mathcal{K}_{\text{after}} = \mathcal{K}_1^0(\text{orig}) + \mathcal{K}_1^0(\text{reg}). \tag{2.44}$$

Since

$$|K^0\rangle = \frac{1}{\sqrt{2}}\left[|K_1^0\rangle + |K_2^0\rangle\right] \longrightarrow \mathcal{K}_1^0(d=0) = \mathcal{K}_2^0(d=0), \tag{2.45}$$

the above two amplitudes read (see (2.36), (2.39) and (2.43))

$$\mathcal{K}_1^0(\text{orig}) = e^{(ik_1 - 1/2\Lambda_1)(d+l)}\, e^{-l/2\mu}\, \mathcal{K}_1^0(d=0) \tag{2.46}$$

and

$$\begin{aligned}
\mathcal{K}_1^0(\text{reg}) &= \mathfrak{r}\left[1 - e^{[i(k_1-k_2)-1/2\Lambda_1]l}\right]\, e^{ik_2(d+l)}\, e^{-l/2\mu}\, \mathcal{K}_1^0(d=0) \\
&= \mathfrak{r}\, e^{ik_1(d+l)}\left[e^{-i(k_1-k_2)(d+l)} - e^{-i(k_1-k_2)d - l/2\Lambda_1}\right] \\
&\quad \times e^{-l/2\mu}\, \mathcal{K}_1^0(d=0) \\
&= e^{-i\delta_m d/\Lambda_1}\, \mathfrak{r}\left[e^{-i\delta_m l/\Lambda_1} - e^{-l/2\Lambda_1}\right] \\
&\quad \times e^{ik_1(d+l)}\, e^{-l/2\mu}\, \mathcal{K}_1^0(d=0),
\end{aligned} \tag{2.47}$$

where

$$\mathfrak{r} \equiv \frac{iN\lambda_1\Lambda_1\mathfrak{f}_r}{-i\delta_m + 1/2}. \tag{2.48}$$

Hence,

$$\begin{aligned}
\mathcal{K}_{\text{after}} &= \left[e^{-(d+l)/2\Lambda_1} + e^{-i\delta_m d/\Lambda_1}\mathfrak{r}\left(e^{-i\delta_m l/\Lambda_1} - e^{-l/2\Lambda_1}\right)\right] \\
&\quad \times e^{ik_1(d+l) - l/2\mu}\, \mathcal{K}_1^0(d=0).
\end{aligned} \tag{2.49}$$

Defining

$$A(l) \equiv \mathfrak{r}\left[e^{-i\delta_m l/\Lambda_1} - e^{-l/2\Lambda_1}\right] \equiv |A(l)|\, e^{i\,\arg A(l)} \tag{2.50}$$

means the (normalized) intensity of K_1^0 mesons right after the regenerator can be expressed as

$$\frac{|\mathcal{K}_{\text{after}}|^2}{|_{K_1^0}(d+l)} = e^{-(d+l)/\Lambda_1} + |A(l)|^2 + 2\,|A(l)|$$

$$\times e^{-(d+l)/2\Lambda_1} \cos\left[\arg A(l) - \delta_{\text{m}} d/\Lambda_1\right]. \tag{2.51}$$

The last term in (2.51) describes the K_1^0-K_1^0 interference. The difference in phase between $\mathcal{K}_1^0(\text{reg})$ and $\mathcal{K}_1^0(\text{orig})$ depends on the proper time elapsed between production and regeneration because of Δm_k. By altering d, this phase difference can be changed to maximize the destructive interference between the two waves. The K_1^0 intensity will exibit a pronounced minimum when

$$\cos\left[\arg A(l) - \delta_{\text{m}} d/\Lambda_1\right] = -1 \quad \text{i.e.} \quad \arg A(l) - \delta_{\text{m}} d/\Lambda_1 = 180°. \tag{2.52}$$

Since the maximum interference occurs when the amplitudes of the two waves are about equal, l can be proportionately reduced as d is increased to keep the ratio of intensities

$$\frac{\left|\mathcal{K}_1^0(\text{reg})\right|^2}{\left|\mathcal{K}_1^0(\text{orig})\right|^2} = \frac{|A(l)|^2}{e^{-(d+l)/\Lambda_1}} \tag{2.53}$$

close to unity.

In the experiment [12], a K^+ beam of 900 MeV/c at Berkeley was used to produce K^0 mesons via $K^+ n \to K^0 p$ in a copper target (see Fig. 2.9). Spark-chamber pictures of $K_1^0 \to \pi^+\pi^-$ events were taken behind an iron regenerator. By measuring the K_1^0 intensity as a function of $(d+l)/\Lambda_1$ one can deduce both the magnitude and sign of Δm_k, provided the magnitude and phase of $\mathfrak{f}_{\text{r}}$ are known. The latter were determined by the authors in a separate scattering experiment with charged kaons on iron nuclei.[12]

The phase of $A(l)$ is the sum of the four nonzero phases in[13]

$$A(l) = iN\lambda_1 \Lambda_1 \mathfrak{f}_{\text{r}} \frac{e^{-i\delta_{\text{m}} l/\Lambda_1} - e^{-l/2\Lambda_1}}{-i\delta_{\text{m}} + 1/2}. \tag{2.54}$$

Based on reference [12], we obtain[14]

$$\arg(\text{i}) = \arg\left[e^{i\pi/2}\right] = \pi/2 = 90°, \quad \arg(\mathfrak{f}_{\text{r}}) = -150°,$$

---

[12] The imaginary part of $\mathfrak{f}_{\text{r}}$ can be determined from the $K^\pm$ cross-sections on nucleons: $\sigma(K^+ n) = \sigma(K^0 p)$, $\sigma(K^- p) = \sigma(\bar{K}^0 n)$, etc. The real part of $\mathfrak{f}$ is obtained from the interference between Coulomb and nuclear scattering ($\bar{\mathfrak{f}}$ is usually assumed to be purely imaginary). To calculate $\mathfrak{f}_{\text{r}}$ for iron nuclei, optical-model fits to nucleon data are used.

[13] The phase of a complex function $f(z) = f_1(z)f_2(z)f_3(z)\cdots$ is the algebraic sum of the phases of its individual factors: $\arg f(z) = \arg f_1(z) + \arg f_2(z) + \arg f_3(z) + \cdots$. Recall also that $\arg[f_1(z)/f_2(z)] = \arg f_1(z) - \arg f_2(z)$.

[14] The average momentum of the K^0 beam was approximately 760 MeV/c, which corresponds to $\Lambda_1 = 3.98\,\text{cm}$ ($l/\Lambda_1 = 1.9$).

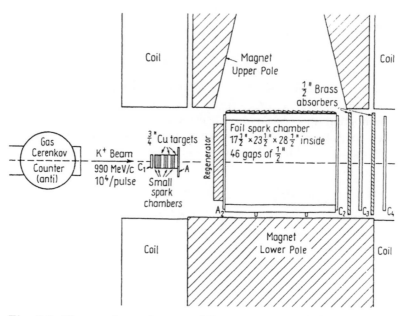

Fig. 2.9. The experimental set-up of W. Mehlhop et al. [12]

$$\arg\left[1/2 - i\delta_{\mathrm{m}}\right]^{-1} = -\arg\left[1/2 - i\delta_{\mathrm{m}}\right] = \arcsin\left[\frac{2\delta_{\mathrm{m}}}{\sqrt{1 + 4\delta_{\mathrm{m}}^2}}\right]$$

$$= \begin{cases} +45^{\circ}, \delta_{\mathrm{m}} = +0.5, \\ -45^{\circ}, \delta_{\mathrm{m}} = -0.5, \end{cases}$$

$$\arg\left[e^{-i\delta_{\mathrm{m}}l/\Lambda_1} - e^{-l/2\Lambda_1}\right] = \arctan\left[\frac{\sin(-\delta_{\mathrm{m}}l/\Lambda_1)}{\cos(-\delta_{\mathrm{m}}l/\Lambda_1) - e^{-l/2\Lambda_1}}\right]$$

$$= \begin{cases} -76^{\circ}, \delta_{\mathrm{m}} = +0.5, \\ +76^{\circ}, \delta_{\mathrm{m}} = -0.5. \end{cases}$$

Hence, $\arg A(l) = -60^{\circ} \mp 31^{\circ}$ for $\delta_{\mathrm{m}} = \pm 0.5$ (see Fig. 2.10). The minimum is given by (2.52), which yields $d/\Lambda_1 = 3.1(\delta_{\mathrm{m}} = +0.5)$ and $7.3(\delta_{\mathrm{m}} = -0.5)$. The $K_1^0 \to \pi^+\pi^-$ decay intensity measured as a function of $(d + l)/\Lambda_1 = d/\Lambda_1 + 1.9$ is shown in Fig. 2.11. This distribution has a minimum at 5.5 (in units of τ_1), corresponding to $d/\Lambda_1 = 5.5 - 1.9 = 3.6$, which clearly favors the value of 3.1 predicted for $\delta_{\mathrm{m}} = +0.5$. We thus conclude that K_2^0 is heavier than K_1^0.

If one assumes a value for δ_{m}, the experiment can provide the magnitude and phase of f_{r}. Excellent agreement was found between $|f_{\mathrm{r}}|$ and ϕ_{r} obtained this way for $\delta_{\mathrm{m}} = 0.46$ and the corresponding values from the scattering experiment with charged kaons mentioned above.

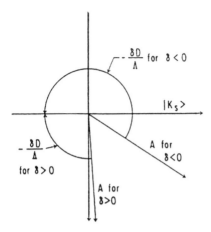

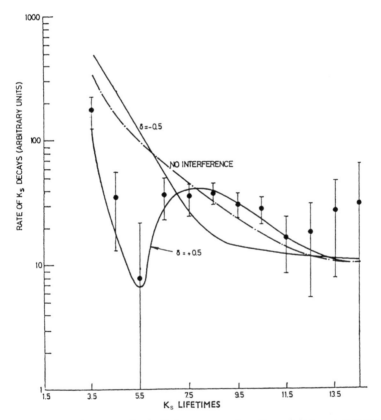

Fig. 2.10. Phase relation between K_1^0(orig) and K_1^0(reg) [12]

Fig. 2.11. The rate of $\pi^+\pi^-$ decays as a function of distance measured by W. Mehlhop et al. [12]

3. *CP* Violation in K^0 Decays

We have assumed up to now that the combined operation of charge conjugation and parity transformation, $\hat{C}\hat{P}$, which turns a particle state into an antiparticle state, is conserved in weak interactions. Considering that parity violation is such a large effect (as we mentioned earlier, all neutrinos are left-handed and all antineutrinos are right-handed), and that both P and C are not conserved in weak interactions (applying $\hat{C}$ to a left-handed neutrino changes it into a left-handed antineutrino), one may wonder if this assumption is justified. It turns out that it is not, as we will now explain.

3.1 Discovery of *CP* Violation

In 1964, J. Christenson, J. Cronin, V. Fitch and R. Turlay[15] detected one 2π event among 500 or so common decays of the long-lived neutral kaon — clear evidence of *CP* violation. Subsequent studies of semileptonic decays of neutral kaons have confirmed this finding (see Sect. 4.1).

Unlike parity violation, which is maximal in weak interactions, *CP* is violated only infinitesimally (at a rate of about 10^{-3}). Moreover, *CP* violation has been observed so far only in the K^0 system. While the nonconservation of parity was readily incorporated into the theory of weak interactions, primarily because the neutrino is adequately described by the Dirac equation for massless particles (the *Weyl equation*), a "natural" way to accomodate *CP* violation has yet to be found.

Invariance under *CP* implies a particle–antiparticle symmetry in nature. As it happens, there is practically no animatter in the universe. From measurements of galactic masses and nucleosynthesis calculations, and from the temperature of the microwave background radiation, the ratio of baryon to photon densities at the present time is found to be

$$\eta_{\text{observed}} = \frac{n_b - \bar{n}_b}{n_\gamma} = \frac{10^{-7}/\text{cm}^3}{400/\text{cm}^3} \approx 10^{-9}.$$

Since the universe is electrically neutral ($n_{e^-} - n_{e^+} = n_p$), there is also an excess of electrons over positrons of about $10^{-7}/\text{cm}^3$. On the other hand the

---

[15] A Princeton University group.

theoretical prediction, based on the assumption of baryon number conservation and initial symmetry between matter and antimatter[16], is $\eta_{\text{th}} \approx 10^{-18}$

As shown in Sect. 4.1, the long-lived neutral kaon decays more often (about 3×10^{-3} times) into a positron than into an electron. If the K_2^0 meson were a pure *CP* eigenstate, and if *CP* were strictly conserved, the two decay modes (4.5), which transform into one another under *CP*, would be equally probable. The nonconservation of *CP* therefore permits particle–antiparticle "discrimination" and thus may be responsible for the observed asymmetry between matter and antimatter in the universe.

The charge asymmetry in the decay of neutral kaons not only distinguishes between matter and antimatter, but also provides an unambigous definition of positive charge: it is the electric charge carried by the lepton preferentially emitted in the decay of K_2^0.

In essence, *CP* violation implies that the laws of nature do make an arbitrary distinction between left and right and between particles and antiparticles. So far, *CP* violation has been observed via the *CP* forbidden decays $K_2^0 \rightarrow \pi^+\pi^-$ [13a], $\pi^0\pi^0$ [13b] and $\pi^+\pi^-\gamma$ [13c], and in the form of charge asymmetry in the semileptonic decays of K_2^0 [26, 27].

In the celebrated *CP*-violation experiment of J. Christenson et al. [13a], a beryllium target was placed in the circulating 30 GeV proton beam of the Brookhaven A.G. Synchrotron. Neutral beams of approximately 1 GeV/c, emitted at 30° to the proton direction, pass through two collimators and a sweeping magnet before entering a plastic "bag" filled with helium gas at atmospheric pressure, placed 17 m from the target. At this point the K_1^0 component has decayed away leaving a pure K_2^0 beam. Pairs of charged particles originating from the (cross-hatched) area inside the helium bag (see Fig. 3.1) are analyzed by two spectrometers consisting of bending magnets and spark chambers, triggered on a coincidence between water Čerenkov and scintillator counters. The helium bag serves to minimize secondary interactions in the decay region ("cheap vacuum"). Spark-chamber photographs were measured on machines equiped with digitized angular encoders.

The rare 2π decays are distinguished from the common semileptonic and 3π decays on the basis of their invariant mass[17] and the direction θ of their resultant momentum vector relative to the incident beam ($\theta \approx 0$ for $K^0 \rightarrow \pi^+\pi^-$). The apparatus was calibrated for 2π events by measuring $K_1^0 \rightarrow \pi^+\pi^-$ decays produced by coherent regeneration in a tungsten regenerator successively placed at intervals of 28 cm along the sensitive decay region. Since the regenerated K_1^0 mesons have the same momentum and direction as the K_2^0 beam, their decays simulate the *CP*-violating $K_2^0 \rightarrow \pi^+\pi^-$ decay. For these measurements a thin anticoincidence counter was placed immediately

---

[16] In the context of the cosmological Big Bang model.

[17] $m_{\pi^+\pi^-} = m_{k^0} = \left[\left(\sum_{i=1}^{2} E_i \right)^2 - \left(\sum_{i=1}^{2} \boldsymbol{P}_i \right)^2 \right]^{1/2}$

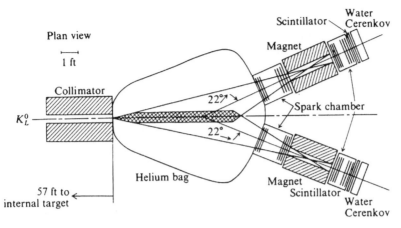

Fig. 3.1. The experimental arrangement of J. Christenson et al. [13a]

behind the regenerator to ensure that K_1^0 mesons decay downstream from it, thereby eliminating neutron-induced background events.

Taking into account the relative detection efficiency for two- and three-body decays, they retained 45 ± 9 events from the helium gas that appeared indentical with those from the coherent regeneration in tungsten in both mass and angular distribution. The total corrected sample of K_2^0 decays was 22700. They concluded that K_2^0 decays to two pions with a branching ratio of

$$\mathcal{R} = \frac{K_2^0 \to \pi^+\pi^-}{K_2^0 \to \text{all}} = (2.0 \pm 0.4) \times 10^{-3}. \tag{3.1}$$

3.2 Phenomenological Implications of $K_2^0 \to 2\pi$

Let us assume — and this is generally believed to be the case — that the weak hamiltonian is not invariant under CP. As a consequence, its eigenstates are not CP eigenstates, but linear superpositions of CP-odd and CP-even components. Since CP violation is so small, we expect the particle eigenstates to differ only slightly from the CP eigenstates K_1^0 and K_2^0. We therefore write the short- and long-lived eigenstates of the weak hamiltonian, $\hat{H}_\text{w}$, as

$$|K_\text{S}^0\rangle \equiv |K_1^0\rangle + \epsilon_1|K_2^0\rangle,$$
$$|K_\text{L}^0\rangle \equiv |K_2^0\rangle + \epsilon_2|K_1^0\rangle, \tag{3.2}$$

i.e.

$$|K_\text{S}^0\rangle = |CP = +1\rangle + \left(10^{-3}\right) \times |CP = -1\rangle,$$
$$|K_\text{L}^0\rangle = |CP = -1\rangle + \left(10^{-3}\right) \times |CP = +1\rangle, \tag{3.3}$$

where $\epsilon_{1,2}$ is generally a complex number. In the above expressions we neglected the normalization factors $1/\sqrt{1 + |\epsilon|^2}$, which are second order in ϵ.

The new states K_S^0 and K_L^0 are clearly not *CP* eigenstates. Moreover, they are not even orthogonal:

$$\langle K_L^0 \mid K_S^0 \rangle = \epsilon_1 + \epsilon_2^*. \tag{3.4}$$

This lack of orthogonality is to be expected since K_S^0 and K_L^0 have the same decay modes. From (1.8), (1.10) and (3.2) it follows (neglecting terms proportional to $\epsilon_1\epsilon_2$) that

$$|K_S^0\rangle = \frac{1}{\sqrt{2}} \left[(1 + \epsilon_1)|K^0\rangle + (1 - \epsilon_1)|\bar{K}^0\rangle \right],$$

$$|K_L^0\rangle = \frac{1}{\sqrt{2}} \left[(1 + \epsilon_2)|K^0\rangle - (1 - \epsilon_2)|\bar{K}^0\rangle \right], \tag{3.5}$$

and

$$|K^0\rangle = \frac{1}{\sqrt{2}} \left[(1 - \epsilon_2)|K_S^0\rangle + (1 - \epsilon_1)|K_L^0\rangle \right],$$

$$|\bar{K}^0\rangle = \frac{1}{\sqrt{2}} \left[(1 + \epsilon_2)|K_S^0\rangle - (1 + \epsilon_1)|K_L^0\rangle \right]. \tag{3.6}$$

If we assume that weak interactions are invariant under the combined operation of charge conjugation, $\hat{C}$, parity transformation, $\hat{P}$, and time reversal, $\hat{T}$, then *CPT* invariance means

$$\epsilon_1 = \epsilon_2 \equiv \epsilon, \tag{3.7}$$

where ϵ is a measure of nature's deviation from perfect *CP* invariance. The above assumption is in accordance with the *CPT theorem*, which states that any quantum theory that is based on relativistic invariance and locality is automatically invariant under $\hat{C}\hat{P}\hat{T}$. An important consequence of this theorem is that a particle and its antiparticle must have the same mass, decay lifetime and magnetic moment.

To prove (3.7), consider a transition $\mathcal{I}(\boldsymbol{p}_i, \boldsymbol{s}_i) \to \mathcal{F}(\boldsymbol{p}_f, \boldsymbol{s}_f)$, where $\mathcal{I}$ stands for one or more particles in the initial state and $\mathcal{F}$ for particles in the final state; $\boldsymbol{p}$ and $\boldsymbol{s}$ are the corresponding momenta and spins, respectively. This transition is described by the matrix element

$$\mathcal{A}_{fi} = \langle \mathcal{F} \mid \hat{A} \mid \mathcal{I} \rangle. \tag{3.8}$$

The operation of $\hat{C}$ flips the signs of *internal charges*, such as the electric charge, baryon number, etc., but spins and momenta are not affected; $\hat{P}$ reverses 3-momenta; under $\hat{T}$, initial and final states are interchanged and spins and momenta are reversed. The effect of the combined $\hat{C}\hat{P}\hat{T}$ operation is thus

$$\langle \mathcal{F} \mid \hat{A} \mid \mathcal{I} \rangle \overset{\hat{C}\hat{P}\hat{T}}{\longrightarrow} \langle \bar{\mathcal{I}}' \mid \hat{A} \mid \bar{\mathcal{F}}' \rangle, \tag{3.9}$$

where the overbar denotes antiparticles and the prime means that spins are reversed. Therefore, assuming *CPT invariance* means

$$\langle \mathcal{F} \mid \hat{A} \mid \mathcal{I} \rangle = \langle \bar{\mathcal{I}}' \mid \hat{A} \mid \bar{\mathcal{F}}' \rangle. \tag{3.10}$$

For spinless kaons this implies (*CPT invariance*)

$$\langle K^0 \mid \hat{H} \mid K^0 \rangle = \langle \bar{K}^0 \mid \hat{H} \mid \bar{K}^0 \rangle. \tag{3.11}$$

If we define

$$\alpha \equiv \frac{1 + \epsilon_1}{\sqrt{2}}, \quad \beta \equiv \frac{1 - \epsilon_1}{\sqrt{2}}, \quad \gamma \equiv \frac{1 + \epsilon_2}{\sqrt{2}}, \quad \delta \equiv \frac{1 - \epsilon_2}{\sqrt{2}} \tag{3.12}$$

expressions (3.5) and (3.6) can be written as ($|K_S^0\rangle \equiv K_S^0$, etc.)

$$\begin{aligned} K_S^0 &= \alpha K^0 + \beta \bar{K}^0, \quad K_L^0 = \gamma K^0 - \delta \bar{K}^0 \\ K^0 &= \delta K_S^0 + \beta K_L^0, \quad \bar{K}^0 = \gamma K_S^0 - \alpha K_L^0. \end{aligned} \tag{3.13}$$

By using (2.9) and (2.10), the time development of the K^0 state is given by

$$K^0 = \delta K_S^0 \, \mathrm{e}^{\phi_S} + \beta K_L^0 \, \mathrm{e}^{\phi_L} = \delta \left(\alpha K^0 + \beta \bar{K}^0 \right) \mathrm{e}^{\phi_S} + \beta \left(\gamma K^0 - \delta \bar{K}^0 \right) \mathrm{e}^{\phi_L},$$

i.e.

$$K^0 = \left(\delta\alpha \, \mathrm{e}^{\phi_S} + \beta\gamma \, \mathrm{e}^{\phi_L} \right) K^0 + \left(\delta\beta \, \mathrm{e}^{\phi_S} - \beta\delta \, \mathrm{e}^{\phi_L} \right) \bar{K}^0. \tag{3.14}$$

Similarly,

$$\bar{K}^0 = \left(\gamma\alpha \, \mathrm{e}^{\phi_S} - \alpha\gamma \, \mathrm{e}^{\phi_L} \right) K^0 + \left(\gamma\beta \, \mathrm{e}^{\phi_S} + \alpha\delta \, \mathrm{e}^{\phi_L} \right) \bar{K}^0. \tag{3.15}$$

From (3.11) and (3.14) and (3.15) it follows that

$$\delta\alpha \, \mathrm{e}^{\phi_S} + \beta\gamma \, \mathrm{e}^{\phi_L} = \gamma\beta \, \mathrm{e}^{\phi_S} + \alpha\delta \, \mathrm{e}^{\phi_L} \longrightarrow \delta\alpha = \gamma\beta.$$

Therefore,

$$\frac{1 - \epsilon_2}{1 + \epsilon_2} = \frac{1 - \epsilon_1}{1 + \epsilon_1} \longrightarrow \epsilon_1 = \epsilon_2 \equiv \epsilon.$$

By assuming *CPT* invariance, expressions (3.5) and (3.6) become

$$\begin{aligned} |K_S^0\rangle &= f(\epsilon) \left[(1 + \epsilon)|K^0\rangle + (1 - \epsilon)|\bar{K}^0\rangle \right], \\ |K_L^0\rangle &= f(\epsilon) \left[(1 + \epsilon)|K^0\rangle - (1 - \epsilon)|\bar{K}^0\rangle \right] \end{aligned} \tag{3.16}$$

and

$$\begin{aligned} |K^0\rangle &= f(\epsilon) \left[(1 - \epsilon)|K_S^0\rangle + (1 - \epsilon)|K_L^0\rangle \right], \\ |\bar{K}^0\rangle &= f(\epsilon) \left[(1 + \epsilon)|K_S^0\rangle - (1 + \epsilon)|K_L^0\rangle \right], \end{aligned} \tag{3.17}$$

where we included the normalization factor

$$f(\epsilon) = \frac{1}{\sqrt{2(1 + |\epsilon|^2)}}. \tag{3.18}$$

CP

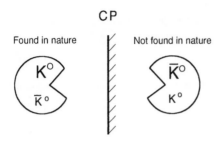

Found in nature | Not found in nature

Fig. 3.2. K_L^0 is a superposion of K^0 and $\bar{K}^0$, with a slightly larger amplitude for K^0. This violates *CP* symmetry

The *CP* nonsymmetric state $|K_L^0\rangle$ is illustrated in Fig. 3.2. Expression (3.4) now reads

$$\langle K_L^0 \mid K_S^0\rangle = \frac{\epsilon + \epsilon^*}{1 + |\epsilon|^2} = \frac{2\mathrm{Re}\,\epsilon}{1 + |\epsilon|^2} = \langle K_S^0 \mid K_L^0\rangle, \; CPT \; invariance. \; (3.19)$$

To relate a possible *CPT*-violating mass difference $m_{k^0} - m_{\bar{k}^0}$ to the measurable quantity $\Delta m_{L,S}$, we consider (3.14), (3.15) in a small time interval Δt. In this case

$$\mathrm{e}^{-(\mathrm{i} m_{S,L} + \Gamma_{S,L}/2)t} \equiv \mathrm{e}^{-\mathrm{i} \mathcal{M}_{S,L} t} \xrightarrow{t \approx 0} 1 - \mathrm{i} \mathcal{M}_{S,L} \Delta t. \quad (3.20)$$

Hence,

$$K^0 \to K^0 - \mathrm{i}\Delta t \left[(\delta \alpha \mathcal{M}_S + \beta \gamma \mathcal{M}_L) K^0 + \beta \delta (\mathcal{M}_S - \mathcal{M}_L) \bar{K}^0 \right],$$
$$\bar{K}^0 \to \bar{K}^0 - \mathrm{i}\Delta t \left[\gamma \alpha (\mathcal{M}_S - \mathcal{M}_L) K^0 + (\gamma \beta \mathcal{M}_S + \alpha \delta \mathcal{M}_L) \bar{K}^0 \right], \quad (3.21)$$

where we used

$$(\delta \alpha + \beta \gamma) K^0 = (1 - \epsilon_1 \epsilon_2) K^0 \approx K^0, \quad \text{etc.}$$

We can express (3.21) as

$$\mathrm{i}\frac{\mathrm{d}\Psi_{\mathrm{vac}}}{\mathrm{d}t_{\mathrm{p}}} =$$

$$\begin{pmatrix} \frac{1}{2}(\mathcal{M}_S + \mathcal{M}_L) + (\mathcal{M}_S - \mathcal{M}_L)\bar{\epsilon} & \frac{1}{2}(\mathcal{M}_S - \mathcal{M}_L)(1 + 2\tilde{\epsilon}) \\ \frac{1}{2}(\mathcal{M}_S - \mathcal{M}_L)(1 - 2\tilde{\epsilon}) & \frac{1}{2}(\mathcal{M}_S + \mathcal{M}_L) - (\mathcal{M}_S - \mathcal{M}_L)\bar{\epsilon} \end{pmatrix} \Psi_{\mathrm{vac}}, \quad (3.22)$$

i.e.

$$\mathrm{i}\frac{\mathrm{d}\Psi_{\mathrm{vac}}}{\mathrm{d}t_{\mathrm{p}}} \equiv \begin{pmatrix} \mathcal{M}_{11} & \mathcal{M}_{12} \\ \mathcal{M}_{21} & \mathcal{M}_{22} \end{pmatrix} \Psi_{\mathrm{vac}} \quad (3.23)$$

with

$$\bar{\epsilon} \equiv \frac{\epsilon_1 - \epsilon_2}{2}, \quad \tilde{\epsilon} \equiv \frac{\epsilon_1 + \epsilon_2}{2}. \quad (3.24)$$

Now,

$$\mathcal{M}_{22} - \mathcal{M}_{11} = (\mathcal{M}_L - \mathcal{M}_S)(\epsilon_1 - \epsilon_2)$$
$$= \left[(m_L - m_S) + i \left(\Gamma_S - \Gamma_L \right)/2 \right](\epsilon_1 - \epsilon_2)$$
$$\approx (m_L - m_S)(\epsilon_1 - \epsilon_2)(1 + \mathrm{i})$$

since $\Delta m_{\mathrm{L,S}} \approx \Gamma_{\mathrm{S}}/2$ (see (1.43)) and $\Gamma_{\mathrm{S}} \gg \Gamma_{\mathrm{L}}$. Therefore,

$$\mathrm{Re}\ (\mathcal{M}_{22} - \mathcal{M}_{11}) \equiv m_{k^0} - m_{\bar{k}^0}$$
$$= (m_{\mathrm{L}} - m_{\mathrm{S}})\,(\epsilon_1 - \epsilon_2) \leq 3.5 \times 10^{-9}\,\mathrm{eV}, \qquad (3.25)$$

which sets the upper limit on a possible *CPT*-violating K^0-$\bar{K}^0$ mass difference. Experimentally [32, 36],

$$\frac{m_{k^0} - m_{\bar{k}^0}}{m_{k^0}} < 1.3 \times 10^{-18}. \qquad (3.26)$$

The above results can be summarized in the following way. According to (3.24),

$$2\tilde{\epsilon} = \epsilon_1 + \epsilon_2, \quad 2\bar{\epsilon} = \epsilon_1 - \epsilon_2 \longrightarrow \epsilon_1 = \tilde{\epsilon} + \bar{\epsilon}, \quad \epsilon_2^* = \tilde{\epsilon}^* - \bar{\epsilon}^*,$$

which we use to express (3.4) as

$$\langle K_{\mathrm{L}}^0 \mid K_{\mathrm{S}}^0 \rangle = \epsilon_1 + \epsilon_2^* = 2\,(\mathrm{Re}\ \tilde{\epsilon} + i\,\mathrm{Im}\ \bar{\epsilon}). \qquad (3.27)$$

- If *CPT* is conserved (and consequently *T* is violated), then according to (3.7) and (3.11)

$$\mathcal{M}_{11} = \mathcal{M}_{22}, \quad \langle K_{\mathrm{L}}^0 \mid K_{\mathrm{S}}^0 \rangle = 2\mathrm{Re}\ \epsilon \ (\epsilon_1 = \epsilon_2), \quad CPT\ invariance. \quad (3.28)$$

- If, on the other hand, *T* is conserved (which means that *CPT* is violated), then

$$\mathcal{M}_{12} = \mathcal{M}_{21}, \quad \langle K_{\mathrm{L}}^0 \mid K_{\mathrm{S}}^0 \rangle = 2i\,\mathrm{Im}\ \bar{\epsilon} \ (\epsilon_1 = -\epsilon_2), \quad T\ invariance. \quad (3.29)$$

because time reversal interchanges indices of the initial and final states.

Ignoring the small *CP* violation, we find that

$$m_{\mathrm{L}} - m_{\mathrm{S}} = -\,(\mathcal{M}_{12} + \mathcal{M}_{12}^*) = -2\mathrm{Re}\ M_{12}. \qquad (3.30)$$

Recall that the mass and decay matrices are both hermitian:

$$M_{21} = M_{12}^*, \quad \Gamma_{21} = \Gamma_{12}^*. \qquad (3.31)$$

3.3 Unitarity, *CPT* Invariance and *T* Violation

The observed *CP* violation in the neutral kaon decay implies that either *T* is violated but *CPT* is conserved, or *CPT* is violated but *T* is conserved, or both *T* and *CPT* are violated. To test these possibilities we will rely on the *unitarity relations*, first derived by J. Bell and J. Steinberger [14a]. These relations are direct consequences of probability conservation, as shown in what follows.

The time development of the $(K_{\mathrm{S}}^0, K_{\mathrm{L}}^0)$ system is given by

$$|\Psi_{\mathrm{S,L}}\rangle = a_{\mathrm{S}}\,\mathrm{e}^{-\mathrm{i}\mathcal{M}_{\mathrm{S}}t}|K_{\mathrm{S}}^0\rangle + a_{\mathrm{L}}\,\mathrm{e}^{-\mathrm{i}\mathcal{M}_{\mathrm{L}}t}|K_{\mathrm{L}}^0\rangle \qquad (3.32)$$

with $\mathcal{M}_{\text{S,L}}$ as defined in (3.20). The states $|K_{\text{S}}^0\rangle$ and $|K_{\text{L}}^0\rangle$ are not orthogonal due to *CP* violation. We can thus write

$$
\begin{aligned}
\langle \Psi_{\text{S,L}} \mid \Psi_{\text{S,L}} \rangle = {} & |a_{\text{S}}|^2 \, e^{-\Gamma_{\text{S}} t} + |a_{\text{L}}|^2 \, e^{-\Gamma_{\text{L}} t} \\
& + a_{\text{S}}^* a_{\text{L}} \, e^{i(\mathcal{M}_{\text{S}}^* - \mathcal{M}_{\text{L}})t} \langle K_{\text{S}}^0 \mid K_{\text{L}}^0 \rangle \\
& + a_{\text{L}}^* a_{\text{S}} \, e^{i(\mathcal{M}_{\text{L}}^* - \mathcal{M}_{\text{S}})t} \langle K_{\text{L}}^0 \mid K_{\text{S}}^0 \rangle .
\end{aligned}
\tag{3.33}
$$

At $t \to 0$, the decrease in the norm of $|\Psi_{\text{S,L}}\rangle$ can be expressed as

$$
\begin{aligned}
-\frac{\mathrm{d}}{\mathrm{d}t} \langle \Psi_{\text{S,L}} \mid \Psi_{\text{S,L}} \rangle = {} & |a_{\text{S}}|^2 \Gamma_{\text{S}} + |a_{\text{L}}|^2 \Gamma_{\text{L}} \\
& - \mathrm{i} \left(\mathcal{M}_{\text{S}}^* - \mathcal{M}_{\text{L}} \right) a_{\text{S}}^* a_{\text{L}} \langle K_{\text{S}}^0 \mid K_{\text{L}}^0 \rangle \\
& - \mathrm{i} \left(\mathcal{M}_{\text{L}}^* - \mathcal{M}_{\text{S}} \right) a_{\text{L}}^* a_{\text{S}} \langle K_{\text{L}}^0 \mid K_{\text{S}}^0 \rangle .
\end{aligned}
\tag{3.34}
$$

Since the kaon decay results in the appearance of the decay product, conservation of probability requires (3.34) to be compensated by the total transition rate

$$
\begin{aligned}
\sum_f \left| \langle f \mid \hat{H}_{\text{w}} \mid \Psi_{\text{S,L}} \rangle \right|^2 = {} & |a_{\text{S}}|^2 \sum_f |A_{\text{S}}|^2 + |a_{\text{L}}|^2 \sum_f |A_{\text{L}}|^2 \\
& + a_{\text{S}}^* a_{\text{L}} \sum_f A_{\text{S}}^* A_{\text{L}} + a_{\text{L}}^* a_{\text{S}} \sum_f A_{\text{L}}^* A_{\text{S}},
\end{aligned}
\tag{3.35}
$$

where the transition amplitudes are

$$
A_{\text{S,L}} \equiv \langle f \mid \hat{H}_{\text{w}} \mid K_{\text{S,L}}^0 \rangle .
\tag{3.36}
$$

Equating coefficients in (3.34) and (3.35) yields the Bell–Steinberger *unitarity relations*

$$
\begin{aligned}
\Gamma_{\text{S}} &= \sum_f |\langle f \mid \hat{H}_{\text{w}} \mid K_{\text{S}}^0 \rangle|^2 \equiv \sum_f |A_{\text{S}}|^2, \\
\Gamma_{\text{L}} &= \sum_f |\langle f \mid \hat{H}_{\text{w}} \mid K_{\text{L}}^0 \rangle|^2 \equiv \sum_f |A_{\text{L}}|^2, \\
-\mathrm{i} \left(\mathcal{M}_{\text{S}}^* - \mathcal{M}_{\text{L}} \right) \langle K_{\text{S}}^0 \mid K_{\text{L}}^0 \rangle &= \sum_f \langle f \mid \hat{H}_{\text{w}} \mid K_{\text{S}}^0 \rangle^* \langle f \mid \hat{H}_{\text{w}} \mid K_{\text{L}}^0 \rangle \\
&\equiv \sum_f A_{\text{S}}^* A_{\text{L}}, \\
-\mathrm{i} \left(\mathcal{M}_{\text{L}}^* - \mathcal{M}_{\text{S}} \right) \langle K_{\text{L}}^0 \mid K_{\text{S}}^0 \rangle &= \sum_f \langle f \mid \hat{H}_{\text{w}} \mid K_{\text{L}}^0 \rangle^* \langle f \mid \hat{H}_{\text{w}} \mid K_{\text{S}}^0 \rangle \\
&\equiv \sum_f A_{\text{L}}^* A_{\text{S}}.
\end{aligned}
\tag{3.37}
$$

Note that in deriving (3.37) we set the same transition rate for different final states. This point will be addressed again shortly.

By using the Schwartz inequality

$$\left| \sum A_{\mathrm{L}}^* A_{\mathrm{S}} \right|^2 \leq \sum |A_{\mathrm{L}}|^2 \cdot |A_{\mathrm{S}}|^2 \tag{3.38}$$

it follows from expressions (3.37) that

$$\left[\left(\frac{\Gamma_{\mathrm{S}} + \Gamma_{\mathrm{L}}}{2} \right)^2 + (m_{\mathrm{L}} - m_{\mathrm{S}})^2 \right] |\langle K_{\mathrm{S}}^0 \mid K_{\mathrm{L}}^0 \rangle|^2 \leq \Gamma_{\mathrm{S}} \Gamma_{\mathrm{L}}, \tag{3.39}$$

i.e.

$$\langle K_{\mathrm{S}}^0 \mid K_{\mathrm{L}}^0 \rangle \leq 0.06, \tag{3.40}$$

which confirms that $|K_{\mathrm{S}}^0\rangle$ and $|K_{\mathrm{L}}^0\rangle$ are nearly orthogonal.

To test *CPT* and *T* invariance we will employ the fourth *unitarity relation*

$$\left[-\mathrm{i}(m_{\mathrm{L}} - m_{\mathrm{S}}) + \frac{\Gamma_{\mathrm{L}} + \Gamma_{\mathrm{S}}}{2} \right] \langle K_{\mathrm{L}}^0 \mid K_{\mathrm{S}}^0 \rangle$$

$$= \sum_f \langle f \mid \hat{H}_{\mathrm{w}} \mid K_{\mathrm{L}}^0 \rangle^* \langle f \mid \hat{H}_{\mathrm{w}} \mid K_{\mathrm{S}}^0 \rangle. \tag{3.41}$$

Both sides of this equation violate *CP* symmetry since (a) $\langle K_{\mathrm{L}}^0 \mid K_{\mathrm{S}}^0 \rangle \neq 0$, and (b) the final states of K_{S}^0 and K_{L}^0 could not otherwise be the same. To simplify (3.41), we define the following *CP*-violating amplitude ratios

$$\eta_f \equiv \frac{\langle f \mid \hat{H}_{\mathrm{w}} \mid K_{\mathrm{L}}^0 \rangle}{\langle f \mid \hat{H}_{\mathrm{w}} \mid K_{\mathrm{S}}^0 \rangle} \quad (CP = +1),$$

$$\eta_f^* \equiv \frac{\langle f \mid \hat{H}_{\mathrm{w}} \mid K_{\mathrm{S}}^0 \rangle}{\langle f \mid \hat{H}_{\mathrm{w}} \mid K_{\mathrm{L}}^0 \rangle} \quad (CP = -1). \tag{3.42}$$

Expression (3.41) now reads

$$\left[-\mathrm{i}\Delta m_{\mathrm{L,S}} + \frac{\Gamma_{\mathrm{L}} + \Gamma_{\mathrm{S}}}{2} \right] \langle K_{\mathrm{L}}^0 \mid K_{\mathrm{S}}^0 \rangle = \sum_f \eta_f^* \Gamma_f, \tag{3.43}$$

where Γ_f is the transition rate to a specific final state, or a group of states. This reflects the fact that various decay modes of K_{S}^0 and K_{L}^0 have different decay rates.

Experimentally, the main three-body decay modes have small decay rates compared with Γ_{S} (see, for example, [15, 21]):

$$\Gamma_{\mathrm{S}}(\text{semileptonic}) \approx \Gamma_{\mathrm{L}}(\text{semileptonic}) \approx \Gamma_{\mathrm{L}}(3\pi) \approx 10^{-3} \Gamma_{\mathrm{S}}. \tag{3.44}$$

The $K_{\mathrm{S}}^0 \to 3\pi$ decays are not only forbidden by *CP* but also suppressed by the phase-space volume[18]; hence $\Gamma_{\mathrm{S}}(3\pi) \ll 10^{-3} \Gamma_{\mathrm{S}}$. The radiative decay mode $K^0 \to \pi^+ \pi^- \gamma$ can also be ignored compared with $K_{\mathrm{S}}^0 \to 2\pi$ [13c].

---

[18] As shown in Appendix A, the $\pi^+ \pi^- \pi^0$ final state is not a *CP* eigenstate. The K_{S}^0 may decay into the kinematics-suppressed but *CP*-allowed final state with $\ell = 1$ and $CP = +1$, or into the kinematics-favored but *CP*-forbidden state with $\ell = 0$ and $CP = -1$, where ℓ is the relative angular momentum between the π^+ and π^-.

Therefore, the right-hand side of (3.43) is dominated by $K_S^0 \to \pi^+\pi^-, \pi^0\pi^0$ decays, i.e., it is sufficiently accurate to write

$$\sum_f \eta_f^* \Gamma_f \approx \eta_{+-}^* \Gamma_{+-} + \eta_{00}^* \Gamma_{00}, \tag{3.45}$$

where Γ_{+-} and Γ_{00} are the partial decay rates for $K_S^0 \to \pi^+\pi^-$ and $K_S^0 \to \pi^0\pi^0$, repectively. The complex amplitude ratios η_{+-} and η_{00} are defined as

$$\frac{\langle \pi^+\pi^- \mid \hat{H}_w \mid K_L^0 \rangle}{\langle \pi^+\pi^- \mid \hat{H}_w \mid K_S^0 \rangle} \equiv |\eta_{+-}| \, e^{i\phi_{+-}}, \quad \frac{\langle \pi^0\pi^0 \mid \hat{H}_w \mid K_L^0 \rangle}{\langle \pi^0\pi^0 \mid \hat{H}_w \mid K_S^0 \rangle} \equiv |\eta_{00}| \, e^{i\phi_{00}}. \tag{3.46}$$

In the next section it will be shown that

$$\Gamma_{+-} + \Gamma_{00} \approx \frac{2}{3}\Gamma_S + \frac{1}{3}\Gamma_S = \Gamma_S. \tag{3.47}$$

Furthermore, the most accurate measurements of the CP-violating parameters η_{+-} and η_{00} are mutually consistent and yield

$$|\eta_{+-}| \approx |\eta_{00}| \approx 2.3 \times 10^{-3}, \quad \phi_{+-} \approx \phi_{00} \approx 44^0. \tag{3.48}$$

Hence,

$$\eta_{+-}^* \Gamma_{+-} + \eta_{00}^* \Gamma_{00} \equiv \eta_{\pi\pi}^* \Gamma_S, \tag{3.49}$$

and (3.43) simplifies to ($\Gamma_L \ll \Gamma_S$)

$$\left[i\frac{\Delta m_{L,S}}{\Gamma_S} + \frac{1}{2} \right] \langle K_L^0 \mid K_S^0 \rangle^* = \eta_{\pi\pi}. \tag{3.50}$$

Using (3.28), we can rewrite this expression as

$$\left[1 + i\frac{2\Delta m_{L,S}}{\Gamma_S} \right] \mathrm{Re}\, \epsilon = |\eta_{\pi\pi}| \cos\phi_{\pi\pi} \left(1 + i \tan\phi_{\pi\pi} \right), \tag{3.51}$$

and similarly for $\langle K_L^0 \mid K_S^0 \rangle = 2i\,\mathrm{Im}\,\bar{\epsilon}$. These results are summarized in Table 3.1 (see [14b]). It is obvious that experiments strongly favour CPT conservation, whereas T is clearly violated. In what follows we shall thus set $\bar{\epsilon} = 0$.

The foregoing analysis represents one of the most stringent tests of CPT invariance in physics. We hasten to point out that it relies solely on some of

Table 3.1.

CPT conserved	T conserved
$\epsilon_1 = \epsilon_2 \equiv \epsilon \ (\bar{\epsilon} = 0)$	$\epsilon_1 = -\epsilon_2 \ (\bar{\epsilon} = 0)$
$\langle K_L^0 \mid K_S^0 \rangle = 2\mathrm{Re}\,\epsilon$	$\langle K_L^0 \mid K_S^0 \rangle = 2i\,\mathrm{Im}\,\bar{\epsilon}$
$\left[1 + i\frac{2\Delta m_{L,S}}{\Gamma_S} \right] \mathrm{Re}\,\epsilon = \eta_{\pi\pi}$	$\left[\frac{2\Delta m_{L,S}}{\Gamma_S} - i \right] \mathrm{Im}\,\bar{\epsilon} = \eta_{\pi\pi}$
$\phi_{\pi\pi} \approx \arctan\left(\frac{2\Delta m_{L,S}}{\Gamma_S} \right) \approx 43.7^\circ$	$\phi_{\pi\pi} \approx \arctan\left(\frac{-\Gamma_S}{2\Delta m_{L,S}} \right) \approx 133.7^\circ$

the basic principles of quantum mechanics: the principle of superposition of amplitudes and conservation of probability.

The analysis also reveals that, for the first time, we have a physical system which behaves asymmetrically in time as a result of an interaction (the weak decay). It is well known that entropy in a closed system increases with time, but this effect is due to boundary conditions, not specific interactions. Moreover, as Stückelberg showed in 1952 [16], the second law of thermodynamics does not depend on the validity of microscopic reversibility, and is thus decoupled from the question of time invariance.

A direct test of time-reversal asymmetry in the K^0 system is discussed in Sect. 7.1.

3.4 Isospin Analysis of $K_{S,L}^0 \to 2\pi$

"Anyone who has played with these invariances knows that it is an orgy of relative phases."

Abraham Pais, Inward Bound

Before presenting experimental results on CP violation in the K^0 system, it is important to make sure that the reader is familiar with the isospin analysis of $K_{S,L}^0 \to 2\pi$ decays [14d].

Recall that pions and kaons are pseudoscalars (spin-zero bosons), whose isospin assignments are given in Table 1.1. The pions are identified as the three states of an isotopic spin triplet, i.e., the pion has isospin $I_\pi = 1$. The neutral kaons belong to an isotopic spin doublet: $I_k = 1/2$. Since pions and kaons are spinless, angular momentum conservation requires the two pions in the $K_{S,L}^0 \to 2\pi$ decay to be in an S state. By Bose statistics, their total wavefunction must be symmetric under particle interchange. Consequently, the isospin function of the 2π-state must also be symmetric.

The isospin states of π^+, π^- and π^0 read

$$|\pi^+\rangle \equiv |1,1\rangle, \quad |\pi^-\rangle \equiv |1,-1\rangle, \quad |\pi^0\rangle \equiv |1,0\rangle, \tag{3.52}$$

and so the possible total isotopic spins of the combined states $|\pi^+\pi^-\rangle$ and $|\pi^0\pi^0\rangle$ are $I = 0$, 1 or 2, with $I_3 = 0$. In terms of Clebsch–Gordan coefficients, we have

$$|0,0\rangle = \sqrt{\frac{1}{3}} |\pi_1^+\pi_2^-\rangle - \sqrt{\frac{1}{3}} |\pi^0\pi^0\rangle + \sqrt{\frac{1}{3}} |\pi_2^-\pi_1^+\rangle,$$

$$|1,0\rangle = \sqrt{\frac{1}{2}} |\pi_1^+\pi_2^-\rangle - \sqrt{\frac{1}{2}} |\pi_2^-\pi_1^+\rangle, \tag{3.53}$$

$$|2,0\rangle = \sqrt{\frac{1}{6}} |\pi_1^+\pi_2^-\rangle + \sqrt{\frac{2}{3}} |\pi^0\pi^0\rangle + \sqrt{\frac{1}{6}} |\pi_2^-\pi_1^+\rangle,$$

where $|\pi^+\pi^-\rangle \equiv |\pi^+\rangle|\pi^-\rangle$, etc. The $|1,0\rangle$ state is not symmetric under particle interchange because it changes sign for $\pi^+ \leftrightarrow \pi^-$. This leaves us with $|2,0\rangle$ and $|0,0\rangle$ as the only allowed isospin states.

To express the experimentally observed pion states $|\pi^+\pi^-\rangle$ and $|\pi^0\pi^0\rangle$ in terms of the allowed isospin states $|2,0\rangle$ and $|0,0\rangle$, note that $|\pi^+\pi^-\rangle$ is the symmetric combination of the two states $|\pi_1^+\pi_2^-\rangle$ and $|\pi_2^-\pi_1^+\rangle$:

$$|\pi^+\pi^-\rangle = \frac{1}{\sqrt{2}}\left[|\pi_1^+\pi_2^-\rangle + |\pi_2^-\pi_1^+\rangle\right]. \tag{3.54}$$

Inverting the two remaining expressions (3.53) gives

$$\begin{aligned}
|\pi^+\pi^-\rangle &= \sqrt{\frac{1}{3}}\,|2,0\rangle + \sqrt{\frac{2}{3}}\,|0,0\rangle, \\[2mm]
|\pi^0\pi^0\rangle &= \sqrt{\frac{2}{3}}\,|2,0\rangle - \sqrt{\frac{1}{3}}\,|0,0\rangle.
\end{aligned} \tag{3.55}$$

In the isotopic spin space, the $K^0 \to 2\pi$ decay is specified by the following four transition amplitudes

$$\begin{aligned}
\langle\pi\pi, I = 0 \mid \hat{H}_{\mathrm{w}} \mid K^0\rangle, &\quad \langle\pi\pi, I = 2 \mid \hat{H}_{\mathrm{w}} \mid K^0\rangle, \\[2mm]
\langle\pi\pi, I = 0 \mid \hat{H}_{\mathrm{w}} \mid \bar{K}^0\rangle, &\quad \langle\pi\pi, I = 2 \mid \hat{H}_{\mathrm{w}} \mid \bar{K}^0\rangle.
\end{aligned} \tag{3.56}$$

The weak decay amplitudes $K^0 \to f$ and $\bar{K}^0 \to f$ are related by CPT invariance (see (3.9)):

$$\begin{aligned}
\langle\bar{f}' \mid \hat{H}_{\mathrm{w}} \mid \bar{K}^0\rangle \xrightarrow{\hat{C}\hat{P}\hat{T}} \langle K^0 \mid \hat{H}_{\mathrm{w}} \mid f\rangle &\equiv \langle f \mid \hat{H}_{\mathrm{w}}^\dagger \mid K^0\rangle^* \\
&= \langle f \mid \hat{H}_{\mathrm{w}} \mid K^0\rangle^* \tag{3.57}
\end{aligned}$$

because $\hat{H}_{\mathrm{w}}$ is hermitian ($\hat{H}_{\mathrm{w}}^\dagger = \hat{H}_{\mathrm{w}}$) in first-order perturbation theory (see Appendix C).

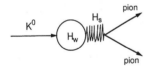

Fig. 3.3. Final-state interaction (H_s) in the decay $K^0 \to 2\pi$

So far we have considered only the weak interaction responsible for the kaon decay. In reality, the final state pions interact strongly with each other (see Fig. 3.3). By resorting again to unitarity and CPT conservation, we can generalize (3.57) to include final state interactions (Watson's theorem; see Appendix C), with the result

$$\langle\bar{f}' \mid \hat{H}_{\mathrm{w}} \mid \bar{K}^0\rangle = \mathrm{e}^{\mathrm{i}2\delta}\,\langle f \mid \hat{H}_{\mathrm{w}} \mid K^0\rangle^*, \tag{3.58}$$

where δ is the scattering phase shift for the states $|f\rangle$ and $|f'\rangle$ at $\sqrt{s} = m_{k^0}$. In the limit of vanishing final-state interactions ($\delta = 0$) this reverts to the old relation between particle and antiparticle transition amplitudes guaranteed by CPT. The above expression can be rewritten as

$$\left[e^{-i\delta}\,\langle \bar{f}' \mid \hat{H}_w \mid K^0 \rangle\right]^* = e^{-i\delta}\,\langle f \mid \hat{H}_w \mid \bar{K}^0 \rangle. \tag{3.59}$$

In view of this result we define the following decay amplitudes

$$A_0 \equiv e^{-i\delta_0}\,\langle \pi\pi,\ I = 0 \mid \hat{H}_w \mid K^0 \rangle,$$
$$A_2 \equiv e^{-i\delta_2}\,\langle \pi\pi,\ I = 2 \mid \hat{H}_w \mid K^0 \rangle, \tag{3.60}$$

where δ_0 and δ_2 are S-wave scattering phases (phase shifts) for $|0,0\rangle$ and $|2,0\rangle$, respectively.[19] These phase shifts are known from the analysis of pion–nucleon scattering, e.g.,

$$\pi^- p \to n\pi^+\pi^-, p\pi^-\pi^0. \tag{3.61}$$

From (3.59) and (3.60) it follows that

$$A_0^* \equiv e^{-i\delta_0}\,\langle \pi\pi,\ I = 0 \mid \hat{H}_w \mid \bar{K}^0 \rangle,$$
$$A_2^* \equiv e^{-i\delta_2}\,\langle \pi\pi,\ I = 2 \mid \hat{H}_w \mid \bar{K}^0 \rangle. \tag{3.62}$$

Taken together, expressions (3.60) and (3.62) read

$$\left.\begin{array}{l} A_\alpha e^{i\delta_\alpha} = \langle I = \alpha \mid \hat{H}_w \mid K^0 \rangle \\ A_\alpha^* e^{i\delta_\alpha} = \langle I = \alpha \mid \hat{H}_w \mid \bar{K}^0 \rangle \end{array}\right\} \alpha = 0 \text{ or } 2. \tag{3.63}$$

We have now established the basic formalism needed to analyze CP violation in $K^0 \to 2\pi$ decays. Using (3.16), (3.55) and (3.63), we obtain

$$\langle \pi^+\pi^- \mid \hat{H}_w \mid K_S^0 \rangle = \sqrt{\frac{2}{3}}\langle 0,0 \mid \hat{H}_w \mid K_S^0 \rangle + \sqrt{\frac{1}{3}}\langle 2,0 \mid \hat{H}_w \mid K_S^0 \rangle$$

$$= f(\epsilon)\sqrt{\frac{1}{3}}\Big\{ \sqrt{2}\,[(1+\epsilon)A_0 e^{i\delta_0} + (1-\epsilon)A_0^* e^{i\delta_0}]$$

$$+ (1+\epsilon)A_2 e^{i\delta_2} + (1-\epsilon)A_2^* e^{i\delta_2}\Big\},$$

i.e.

$$\langle \pi^+\pi^- \mid \hat{H}_w \mid K_S^0 \rangle = f(\epsilon)\sqrt{\frac{4}{3}}\Big\{ \sqrt{2}\,[\text{Re}\,A_0 + i\epsilon\text{Im}\,A_0]\,e^{i\delta_0}$$

$$+ [\text{Re}\,A_2 + i\epsilon\text{Im}\,A_2]\,e^{i\delta_2}\Big\}. \tag{3.64}$$

---

[19] Because pions are present only in the final state they acquire half the usual strong-interaction phase 2δ.

Similarly,

$$\langle \pi^+\pi^- \mid \hat{H}_w \mid K_L^0 \rangle = f(\epsilon)\sqrt{\frac{4}{3}}\Big\{ \sqrt{2}\,[\epsilon \mathrm{Re}\, A_0 + i\, \mathrm{Im}\, A_0]\, \mathrm{e}^{i\delta_0}$$
$$+ [\epsilon \mathrm{Re}\, A_2 + i\, \mathrm{Im}\, A_2]\, \mathrm{e}^{i\delta_2} \Big\}, \tag{3.65}$$

$$\langle \pi^0\pi^0 \mid \hat{H}_w \mid K_S^0 \rangle = f(\epsilon)\sqrt{\frac{4}{3}}\Big\{ -[\mathrm{Re}\, A_0 + i\epsilon \mathrm{Im}\, A_0]\, \mathrm{e}^{i\delta_0}$$
$$+ \sqrt{2}\,[\mathrm{Re}\, A_2 + i\epsilon \mathrm{Im}\, A_2]\, \mathrm{e}^{i\delta_2} \Big\}, \tag{3.66}$$

$$\langle \pi^0\pi^0 \mid \hat{H}_w \mid K_L^0 \rangle = f(\epsilon)\sqrt{\frac{4}{3}}\Big\{ -[\epsilon \mathrm{Re}\, A_0 + i\, \mathrm{Im}\, A_0]\, \mathrm{e}^{i\delta_0}$$
$$+ \sqrt{2}\,[\epsilon \mathrm{Re}\, A_2 + i\, \mathrm{Im}\, A_2]\, \mathrm{e}^{i\delta_2} \Big\}. \tag{3.67}$$

These results can be simplified by observing that the relative contribution of the $I = 2$ decay mode to the $K^0 \rightarrow 2\pi$ transition amplitude is small compared with that of the $I = 0$ mode, as shown below. Since $I_k = 1/2$, the decay $K^0(\bar{K}^0) \rightarrow 2\pi$ can be regarded as a mixture of $\Delta I = 1/2$ and $3/2$ transitions. Experimentally, the branching ratio

$$\frac{\Gamma_{S,+-}}{\Gamma_{S,00}} \equiv \frac{\Gamma(K_S^0 \rightarrow \pi^+\pi^-)}{\Gamma(K_S^0 \rightarrow \pi^0\pi^0)} = \frac{68.6}{31.3} = 2.18 \tag{3.68}$$

is close to that expected for a pure $|0,0\rangle$ final state (see (3.55)):

$$\mathcal{R}_{I=0} = \frac{\left| \langle \pi^+\pi^-; I=0 \mid \hat{H}_w \mid K_S^0 \rangle \right|^2}{\left| \langle \pi^0\pi^0; I=0 \mid \hat{H}_w \mid K_S^0 \rangle \right|^2} = 2. \tag{3.69}$$

This is in agreement with the empirical "$\Delta I = 1/2$ rule", which states that weak amplitudes for processes with $\Delta I = 1/2$ are strongly enhanced compared with those with $\Delta I > 1/2$. For example, the decay $K^+ \rightarrow \pi^+\pi^0$ ($\Delta I = 3/2$) is about 600 times slower than $K_S^0 \rightarrow \pi^+\pi^-$ ($\Delta I = 1/2$). Neglecting a small phase-space correction factor (see (9.156)) and the terms containing ϵ, we have from (3.64) and (3.66) that

$$\frac{\Gamma_{S,+-}}{\Gamma_{S,00}} \equiv \frac{\left| \langle \pi^+\pi^- \mid \hat{H}_w \mid K_S^0 \rangle \right|^2}{\left| \langle \pi^0\pi^0 \mid \hat{H}_w \mid K_S^0 \rangle \right|^2} \approx 2\, \frac{\left| 1 + \frac{1}{\sqrt{2}}\mathrm{Re}\,(A_2/A_0)\, \mathrm{e}^{i(\delta_2-\delta_0)} \right|^2}{\left| 1 - \sqrt{2}\mathrm{Re}\,(A_2/A_0)\, \mathrm{e}^{i(\delta_2-\delta_0)} \right|^2} \tag{3.70}$$

which yields

$$\mathrm{Re}\left(\frac{A_2}{A_0} \right) \approx 0.04 \tag{3.71}$$

for the $\pi\pi$ scattering phase $\pi/2 + \delta_2 - \delta_0 = (37.5 \pm 5)°$.

In what follows we will neglect the terms proportional to $\epsilon \, \mathrm{Re}\,(A_2/A_0)$ and $\epsilon(\mathrm{Im}\,A_2/\mathrm{Re}\,A_0)$. The experimentally observed amplitude ratios (3.46) are then given by

$$
\begin{aligned}
\eta_{+-} &\approx \frac{\epsilon + \mathrm{i}\,\mathrm{Im}\,A_0/\mathrm{Re}\,A_0 + \frac{1}{\sqrt{2}}(\mathrm{i}\,\mathrm{Im}\,A_2/\mathrm{Re}\,A_0)\mathrm{e}^{\mathrm{i}(\delta_2-\delta_0)}}{1 + \frac{1}{\sqrt{2}}\mathrm{Re}\,(A_2/A_0)\mathrm{e}^{\mathrm{i}(\delta_2-\delta_0)}} \\
&\equiv \frac{\varepsilon + \varepsilon_2}{1+\xi} \approx \varepsilon + \varepsilon',
\end{aligned}
\tag{3.72}
$$

$$
\begin{aligned}
\eta_{00} &\approx \frac{\epsilon + \mathrm{i}\,\mathrm{Im}\,A_0/\mathrm{Re}\,A_0 - \sqrt{2}(\mathrm{i}\,\mathrm{Im}\,A_2/\mathrm{Re}\,A_0)\mathrm{e}^{\mathrm{i}(\delta_2-\delta_0)}}{1 - \sqrt{2}\mathrm{Re}\,(A_2/A_0)\mathrm{e}^{\mathrm{i}(\delta_2-\delta_0)}} \\
&\equiv \frac{\varepsilon - 2\varepsilon_2}{1-2\xi} \approx \varepsilon - 2\varepsilon'
\end{aligned}
\tag{3.73}
$$

in any phase convention for which $|\epsilon(\mathrm{Im}\,A_0/\mathrm{Re}\,A_0)| \ll 1$. In (3.72) and (3.73) we defined

$$
\begin{aligned}
\varepsilon &\equiv \epsilon + \frac{\mathrm{i}\,\mathrm{Im}\,A_0}{\mathrm{Re}\,A_0}, \\
\varepsilon_2 &\equiv \frac{\mathrm{i}}{\sqrt{2}}\frac{\mathrm{Im}\,A_2}{\mathrm{Re}\,A_0}\mathrm{e}^{\mathrm{i}(\delta_2-\delta_0)}, \\
\xi &\equiv \frac{1}{\sqrt{2}}\mathrm{Re}\left(\frac{A_2}{A_0}\right)\mathrm{e}^{\mathrm{i}(\delta_2-\delta_0)}
\end{aligned}
\tag{3.74}
$$

and

$$
\varepsilon' \equiv \varepsilon_2 - \frac{\mathrm{i}\,\mathrm{Im}\,A_0}{\mathrm{Re}\,A_0}\,\xi = \frac{\mathrm{i}}{\sqrt{2}}\frac{\mathrm{Re}\,A_2}{\mathrm{Re}\,A_0}\left[\frac{\mathrm{Im}\,A_2}{\mathrm{Re}\,A_2} - \frac{\mathrm{Im}\,A_0}{\mathrm{Re}\,A_0}\right]\mathrm{e}^{\mathrm{i}(\delta_2-\delta_0)}.
\tag{3.75}
$$

Using expressions (3.64)–(3.67) and

$$
\begin{aligned}
\langle 0,0| &= \sqrt{\frac{2}{3}}\,\langle \pi^+\pi^-| - \sqrt{\frac{1}{3}}\,\langle \pi^0\pi^0|, \\
\langle 2,0| &= \sqrt{\frac{1}{3}}\,\langle \pi^+\pi^-| + \sqrt{\frac{2}{3}}\,\langle \pi^0\pi^0|,
\end{aligned}
\tag{3.76}
$$

it is straightforward to show that the parameters ε, ε_2 and ξ can be expressed as ratios of K_S^0 and K_L^0 transition amplitudes to the $I=0$ and $I=2$ states:

$$
\begin{aligned}
\varepsilon &\equiv \frac{\langle 0,0 \mid \hat{H}_\mathrm{w} \mid K_L^0\rangle}{\langle 0,0 \mid \hat{H}_\mathrm{w} \mid K_S^0\rangle}, \\
\varepsilon_2 &\equiv \frac{1}{\sqrt{2}}\frac{\langle 2,0 \mid \hat{H}_\mathrm{w} \mid K_L^0\rangle}{\langle 0,0 \mid \hat{H}_\mathrm{w} \mid K_S^0\rangle}, \\
\xi &\equiv \frac{1}{\sqrt{2}}\frac{\langle 2,0 \mid \hat{H}_\mathrm{w} \mid K_S^0\rangle}{\langle 0,0 \mid \hat{H}_\mathrm{w} \mid K_S^0\rangle}.
\end{aligned}
\tag{3.77}
$$

The phase of ε' is determined from (3.75) (note that $i = e^{i\pi/2}$). Based on (3.81) and (3.82), the phase of $\varepsilon = \epsilon + i \operatorname{Im} A_0/\operatorname{Re} A_0 \equiv \epsilon + i\xi_0$ is, to a good approximation, $\phi_\varepsilon \approx \phi_{\pi\pi} \approx \pi/4$ (see (9.150)):

$$\phi_\varepsilon \approx \arctan\left(\frac{2\Delta m_{\mathrm{L,S}}}{\Gamma_{\mathrm{S}}}\right), \quad \phi_{\varepsilon'} = \pi/2 + \delta_2 - \delta_0. \tag{3.78}$$

We see that η_{+-} and η_{00} do not vanish even without K_{S}^0-K_{L}^0 mixing ($\epsilon = 0$), if either A_0 or A_2 has a nonzero imaginary part. In other words, A_0 and A_2 are real if CP is conserved.

The parameter ε', which affects the $\pi^+\pi^-$ and $\pi^0\pi^0$ channels differently, determines the magnitude of *direct CP* violation in $|\Delta S| = 1$ decays. If CP violation is associated with decay only, then $K_2^0 \to 2\pi$ is possible. Note that ε' is expressed as a difference in CP violation in the $I = 0$ and $I = 2$ decay amplitudes. It vanishes only if A_0 and A_2 have the same phase (see Appendix C and Sect. 9.3).

Adopting the phase convention $\operatorname{Im} A_0 = 0$ eliminates direct CP violation in the $I = 0$ decay mode. The parameter ϵ arises primarily from mixing in the mass matrix, and is thus associated with *indirect CP* violation that mixes CP-even states into K_2^0 ($|\Delta S| = 2$).

To show this, we invert expressions for the off-diagonal matrix elements in (3.22), assuming CPT conservation ($\tilde{\epsilon} = \epsilon$), with the result

$$\epsilon = \frac{\mathcal{M}_{12} - \mathcal{M}_{21}}{2(\mathcal{M}_{\mathrm{S}} - \mathcal{M}_{\mathrm{L}})} = \frac{(\Gamma_{12} - \Gamma_{12}^*) + i(M_{12} - M_{12}^*)}{(\Gamma_{\mathrm{S}} - \Gamma_{\mathrm{L}}) + 2i(m_{\mathrm{S}} - m_{\mathrm{L}})}$$

$$\approx \frac{i \operatorname{Im} \Gamma_{12} - \operatorname{Im} M_{12}}{\Delta m_k(1 - i\, 2\Delta m_k/\Gamma_{\mathrm{S}})}, \tag{3.79}$$

where we used $\mathcal{M}_{ij} = M_{ij} - i\Gamma_{ij}$, $\Gamma_{21} = \Gamma_{12}^*$, $M_{21} = M_{12}^*$ and $2\Delta m_k \approx \Gamma_{\mathrm{S}}$. According to Table 3.1 and (3.72)–(3.74),

$$\operatorname{Re} \epsilon = \frac{\eta_{\pi\pi}}{1 + i\, 2\Delta m_k/\Gamma_{\mathrm{S}}}$$

$$\approx \frac{\epsilon + i\xi_0}{1 + i} \approx i\left(\frac{\operatorname{Im} \Gamma_{12}}{2\Delta m_k} + \frac{\xi_0}{2}\right) + \left(\frac{-\operatorname{Im} M_{12}}{2\Delta m_k} + \frac{\xi_0}{2}\right), \tag{3.80}$$

because $3\eta_{\pi\pi} \approx 2\eta_{+-} + \eta_{00} \approx 3\varepsilon$ and $\varepsilon = \epsilon + i \operatorname{Im} A_0/\operatorname{Re} A_0 \equiv \epsilon + i\xi_0$. Since $\operatorname{Re} \epsilon$ is a real quantity,

$$\frac{\operatorname{Im} \Gamma_{12}}{\Delta m_k} \approx -\xi_0, \quad 2\operatorname{Re} \epsilon \approx \frac{-\operatorname{Im} M_{12}}{\Delta m_k} + \xi_0. \tag{3.81}$$

Now, we can always define the relative phase of K^0 and $\bar{K}^0$ by choosing $\operatorname{Im} A_0 = 0$ or $\operatorname{Im} A_2 = 0$ (see footnote 5). In either case the experimental result $\varepsilon'/\varepsilon \approx 10^{-3}$ (see Sect. 5.1) implies that $|\xi_{0,2}| \ll |\epsilon|$. This shows that ϵ arises primarily from mixing in the mass matrix.

CP violation, as measured by the parameter ϵ, is an effect of the order of 10^{-3} relative to weak transitions: $\operatorname{Im} M_{12} \approx 10^{-8}$ eV compared with $\Delta m_k \approx 3.5 \times 10^{-6}$ eV. Using (3.79) and (3.30), we obtain

$$\epsilon = \frac{\mathrm{i}\,\mathrm{Im}\,\Gamma_{12} - \mathrm{Im}\,M_{12}}{\Delta m_k (1 - \mathrm{i}\tan\phi_{\pi\pi})} \approx \frac{-\mathrm{Im}\,M_{12}}{\sqrt{2}\,\Delta m_k}\,\mathrm{e}^{\mathrm{i}\phi_{\pi\pi}} = \frac{1}{2\sqrt{2}}\frac{\mathrm{Im}\,M_{12}}{\mathrm{Re}\,M_{12}}\,\mathrm{e}^{\mathrm{i}\phi_{\pi\pi}}. \quad (3.82)$$

If M_{12} and Γ_{12} were both real, ϵ would be zero (see (3.79)). Since $|\epsilon|$ is so small,

$$\mathrm{Im}\,M_{12} \ll \mathrm{Re}\,M_{12}, \quad \mathrm{Im}\,\Gamma_{12} \ll \mathrm{Re}\,\Gamma_{12}$$

As promised in the preceding section, we now derive expression (3.47). First note that

$$\langle I = 0 \mid \hat{H}_w \mid K_S^0 \rangle = \frac{1}{\sqrt{2}}\left[\langle I = 0 \mid \hat{H}_w \mid K^0 \rangle + \langle I = 0 \mid \hat{H}_w \mid \bar{K}^0 \rangle \right]$$

$$= \frac{2}{\sqrt{2}}\,\mathrm{Re}\,A_0 \mathrm{e}^{\mathrm{i}\delta_0}. \quad (3.83)$$

We can use the fact that $\mathrm{Re}\,A_2 \ll \mathrm{Re}\,A_0$ to write

$$\Gamma_S \approx \left| \langle I = 0 \mid \hat{H}_w \mid K_S^0 \rangle \right|^2 = 2\,(\mathrm{Re}\,A_0)^2 \quad (3.84)$$

and (neglecting the terms with ϵ and A_2 in (3.64) and (3.66))

$$\Gamma_{S,+-} \equiv \left| \langle \pi^+\pi^- \mid \hat{H}_w \mid K_S^0 \rangle \right|^2 \approx \frac{4}{3}\,(\mathrm{Re}\,A_0)^2,$$

$$\Gamma_{S,00} \equiv \left| \langle \pi^0\pi^0 \mid \hat{H}_w \mid K_S^0 \rangle \right|^2 \approx \frac{2}{3}\,(\mathrm{Re}\,A_0)^2. \quad (3.85)$$

$$(3.86)$$

3.5 K_L^0-K_S^0 Interference as Evidence for CP Violation

After the detection of the $K_L^0 \to \pi^+\pi^-$ decay by the Princeton group in 1964, most physicists were not keen to forsake yet another cherished symmetry principle. They still believed that CP was an exact symmetry of nature which successfully replaced the defunct idea of parity conservation. Various attempts were made to explain the effect without abandoning CP invariance. Among other suggestions, it was proposed that either the observed pions, or their long-lived parent particles, were not the usual mesons; or that, similar to Pauli's neutrino hypothesis, the decay was accompanied by a third particle with small mass and energy.[20]

One can confront these hypotheses by proving that: (a) the mass and lifetime of the particle responsible for the 2π-decay are the same as those of the K^0; (b) the rate of $K_L^0 \to 2\pi$ decays does not depend on how the kaons are produced; and (c) the decay products are similar to pions in every respect.

---

[20] It was even suggested that the Princeton result might be due to regeneration of K_1^0 mesons in a fly trapped in the helium bag! It turns out that if that had been the case the fly would have to be more dense than uranium [92].

Such tests are, in fact, not necessary, for there is a direct proof that the effect is indeed due to *CP* violation. In 1965, V. Fitch and his collaborators observed constructive interference from a coherent beam of K_L^0 and K_S^0 mesons by comparing the decay rates of $K_L^0 \to 2\pi$ in vacuum and in the presence of a diffuse regenerator [17].

In addition to providing conclusive evidence for *CP* violation, the results of their experiment also demonstrated that it is possible to make clear empirical distinction between a world made of matter and that composed of antimatter, as we will now explain.

Suppose that the regenerator in Fitch's experiment is composed of antimatter. In this case the forward scattering amplitudes $\mathfrak{f}$ and $\bar{\mathfrak{f}}$ would be interchanged because strong interactions conserve C invariance. As a consequence, the regeneration amplitude would change sign (see (2.39) and (2.40)):

$$\varrho_c(\text{matter}) \longrightarrow -\varrho_c(\text{antimatter}) \tag{3.87}$$

resulting in destructive interference. Before presenting their results, we will outline the basic idea behind the experiment.

Immediately after the regenerator ($t = 0$), an initially pure K_L^0 beam contains a K_S^0 component (see Fig. 2.7):

$$|\Psi(t = 0)\rangle = |K_L^0(t = 0)\rangle + \varrho_c|K_S^0(t = 0)\rangle.$$

Here we set the K_L^0 amplitude to unity, in which case the amplitude of K_S^0 is just the regeneration amplitude

$$\varrho_c \equiv |\varrho_c|e^{i\phi_\varrho}. \tag{3.88}$$

At a later time t,

$$|\Psi(t)\rangle = e^{-i\mathcal{M}_L t}|K_L^0(t = 0)\rangle + \varrho_c e^{-i\mathcal{M}_S t}|K_S^0(t = 0)\rangle. \tag{3.89}$$

The $\Psi \to \pi^+\pi^-$ amplitude reads

$$\begin{aligned}
\mathsf{A}_{\pi^+\pi^-} &= \langle \pi^+\pi^- \,|\, \hat{H}_w \,|\, \Psi(t)\rangle \\
&= \mathsf{A}_{K_S^0 \to \pi^+\pi^-} \left[\eta_{+-}e^{-i\mathcal{M}_L t} + \varrho_c e^{-i\mathcal{M}_S t} \right]
\end{aligned} \tag{3.90}$$

and the corresponding decay rate (per K_L^0 meson) is

$$\begin{aligned}
\mathsf{I}_{\pi^+\pi^-} = \Gamma_{S,+-} \Big\{ &|\eta_{+-}|^2 e^{-\Gamma_L t} + |\varrho_c|^2 e^{-\Gamma_S t} \\
&+ 2|\eta_{+-}||\varrho_c|e^{-(\Gamma_S + \Gamma_L)t/2} \cos\left[\delta_m t - (\phi_{+-} - \phi_\varrho) \right] \Big\}.
\end{aligned} \tag{3.91}$$

where t is the proper time of decay.

The value of $|\eta_{+-}|^2$ was obtained from the rate of $K_L^0 \to \pi^+\pi^-$ decays without regenerator;[21] that of $|\varrho_c|^2$ was determined by measuring the $\pi^+\pi^-$ decay rate immediately behind a dense regenerator ($\varrho_c \gg \eta_{+-}$), in which case the interference is small (for this measurement a 7.6 cm solid piece of berillium was employed).

---

[21] The result was in excellent agreement with that of Christenson et al. [13a].

The maximum interference occurs when the K_L^0 and K_S^0 amplitudes are about equal. For a berillium regenerator this corresponds to a density of roughly $0.1\,\mathrm{g/cm}^3$, or $N \approx 7 \times 10^{21}$ nuclei/cm^3. To attain low density, they used berillium plates 0.5 mm thick, separated by 1 cm; the whole assembly was 1 m long.

Note that granularity effects in this case are negligible, since the element spacing is small compared with the K_S^0 decay length $\Lambda_S = (p_k/m_k)/\Gamma_S \approx (1.3\,\mathrm{GeV}/0.5\,\mathrm{GeV}) \times 10^{-10}\,\mathrm{s} \approx 2.6 \times 10^{-10} \times 3 \times 10^{10}\,\mathrm{cm} \approx 7.5\,\mathrm{cm}$ and the oscillation length $l_{\mathrm{osc}} \approx 12\Lambda_S \approx 90\,\mathrm{cm}$. This arrangement is, therefore, equivalent to a uniform distribution of the same amount of berillium over l_{osc}.

Since the length of the diffuse regenerator is considerably larger than Λ_S, the coherent regeneration amplitude is independent of $\zeta \equiv l/\Lambda_S$ at distances sufficiently far from the face of the regenerator:

$$\varrho_c \xrightarrow{l \gg \Lambda_S} \frac{iN_0 \lambda_S \Lambda_S \mathfrak{f}_r}{-i\delta_m + 1/2} \equiv \varrho_0. \tag{3.92}$$

That is to say, at large l the number of regenerated K_S^0 mesons is equal to the number of those which decay ($\mathsf{A}_{K_S^0} = \mathrm{const}$). The interference experiments are always performed over a time scale $t \ll \tau_L$, i.e., over a decay length $l_{\mathrm{dec}} \ll \Lambda_L$, which means that the K_L^0 decays can be neglected ($\mathsf{A}_{K_L^0} = \mathrm{const}$). Expression (3.91) for the $\pi^+\pi^-$ decay rate thus simplifies to

$$\mathsf{I}_d \equiv \Gamma_{S,+-} \left\{ |\eta_{+-}|^2 + |\varrho_0|^2 + 2|\eta_{+-}||\varrho_0| \cos(\phi_{+-} - \phi_\varrho) \right\}. \tag{3.93}$$

From (3.92) and (2.42) it folows that

$$|\varrho_c|^2 = |\varrho_0|^2 \left(\frac{N}{N_0}\right)^2 \left[1 + e^{-\zeta_r} - 2e^{-\zeta_r/2} \cos(\delta_m \zeta_r)\right], \tag{3.94}$$

where N and N_0 are the nuclear densities of the solid and the diffuse regenerator, respectively.

The experiment was performed at the Brookhaven AGS. The K_L^0 beam was similar to that used in the original CP violation experiment [13a]. The experimental set-up of Fitch et al. is shown in Fig. 3.4. A particularly useful feature of the detector is that the pion trajectories intersect at a point which lies approximately on the extrapolated line of flight of the kaons. Spark chambers were used as tracking devices.

To retain the applicability of (3.92), they considered only those events that originated in the last 45 cm of the diffuse regenerator. The regeneration amplitude ϱ_0 was determined from (3.94) by measuring $|\varrho_c|^2$ as described above. The intensity I_d was measured over the same kaon decay length l_{dec} as the free-space intensity $\mathsf{I}_a \equiv |\eta_{+-}|^2$. Using (3.93), the phase angle between η_{+-} and ϱ_0 is given by

$$\cos(\phi_{+-} - \phi_\varrho) = \frac{\mathsf{I}_d - \mathsf{I}_a - \mathsf{I}_p}{2\sqrt{\mathsf{I}_a \mathsf{I}_p}} \tag{3.95}$$

with

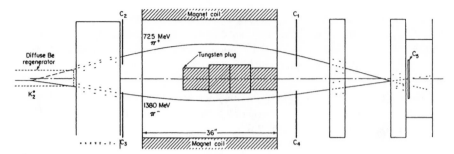

Fig. 3.4. The experimental set-up used by V. Fitch et al. [17]

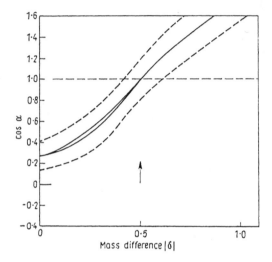

Fig. 3.5. $\cos(\phi_{+-} - \phi_{\varrho})$ versus Δm_k calculated from the [17] data

$$l_{\mathrm{p}} \equiv |\varrho_0|^2 \frac{l_{\mathrm{dec}}}{\Lambda_{\mathrm{S}}}, \tag{3.96}$$

where $l_{dec}/\Lambda_{\mathrm{S}}$ is the decay length scale factor.

The data were corrected for attenuation in berillium and for interference in the regeneration in solid Be. The rate of $\pi^+\pi^-$ decays was found to be about four times the rate without regenerator — a clear indication of *constructive interference* between the $K_{\mathrm{L}}^0 \to \pi^+\pi^-$ and $K_{\mathrm{S}}^0 \to \pi^+\pi^-$ decays.

According to (3.92), ϱ_0 depends on the value of $\delta_{\mathrm{m}} \equiv \Delta m_{\mathrm{L,S}}$. Consequently, $\cos(\phi_{+-} - \phi_{\varrho})$ calculated from the data also is a function of δ_m, as shown in Fig. 3.5. Clearly, the data strongly rejects values of $|\delta_m| > 0.8$.[22]

In their second paper, Fitch et al. reported the $\pi^+\pi^-$ yield as a function of the diffuse regenerator amplitude $A_{\mathrm{r}} \equiv \varrho_0$. Measurements were made at densities of $0, 0.05, 0.1$ and $0.4\,\mathrm{g/cm^3}$. As explained earlier, in a world made

---

[22] The correction for interference in solid Be causes the curve in Fig. 3.5 to be double valued.

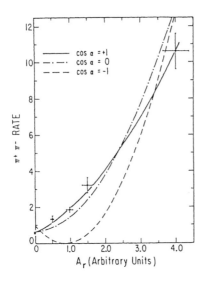

Fig. 3.6. Rate of $\pi^+\pi^-$ decays as a function of the regenerator amplitude A_r [17]

of antimatter an experimenter would observe destructive interference, similar to the dashed line in Fig. 3.6. Note that the result shown in the figure is independent of the sign and magnitude of $\Delta m_{L,S}$.

Interference effects between K_L^0 and K_S^0 mesons can also be studied by measuring the time dependence of the $\pi^+\pi^-$ decay intensity following a thick regenerator. Several such experiments have been performed (e.g., [18, 19]). The diffuse and solid regenerator techniques are nicely summarized by J.-M. Gaillard in [20]. Each group referenced above used different regenerator thicknesses, resulting in increased sensitivity for interference effects at different distances from the regenerator. Both experiments were performed at the CERN Proton Synchrotron (PS).

Figure 3.7 shows the measurement of C. Alff-Steinberger et al., who used copper as the regenerating material (M. Bott-Bodenhausen et al. employed carbon plates). The theoretical expectations, expression (3.91), for different regenerator densities were fitted to the data to obtain the best values of $|\varrho_c|$, $|\eta_{+-}|$, $\Delta m_{L,S}$ and $\phi \equiv \phi_{+-} - \phi_\varrho$ (Γ_S was known to an adequate precision). The distributions corresponding to the best fit solutions are also presented in Fig. 3.7.

No acceptable fit was obtained without the interference term. The latter was extracted from the data for each distribution in Fig. 3.7 by subtracting the quadratic terms according to the fitted parameters $|\varrho_c|$ and $|\eta_{+-}|$, and then dividing by $2|\varrho_c||\eta_{+-}|\exp^{-(\Gamma_S+\Gamma_L)t/2}$. The results from both groups are shown in Fig. 3.8. The oscillatory behavior of the interference term provides a precise measurement of the mass difference $|\Delta m_{L,S}|$ and the phase difference ϕ (remember, $m_L - m_S > 0$, as demonstrated by W. Mehlhop et al. [12]). Since we will describe shortly an experiment that measured the phase difference

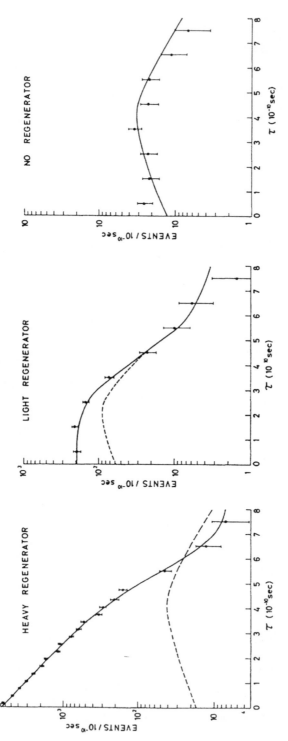

Fig. 3.7. Event rates measured by C. Alff-Steinberger et al. [18]

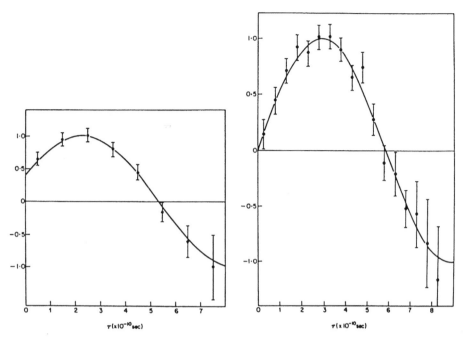

Fig. 3.8. Interference terms measured by C. Alff-Steinberger et al. (*left*) and M. Bott-Bodenhausen et al. (*right*)

$\phi = \phi_{+-} - \phi_\varrho$ and, concurrently, the regeneration phase ϕ_ϱ much more precisely than the CERN groups did, we do not present the values of their fitted parameters.

4. Interference in Semileptonic and Pionic Decay Modes

4.1 Semileptonic Decays of Neutral Kaons

In our discussion of the K^0 system so far we have associated the $(K^0, \bar{K}^0)$ doublet with strong interactions and the (K_1^0, K_2^0) doublet with weak decays. This classification is, in fact, not entirely correct. The reason is that the semileptonic weak decays of neutral kaons (see [21]),

$$\text{Neutral kaon} \longrightarrow e^{\pm} \pi^{\mp} \nu_e, \quad \mu^{\pm} \pi^{\mp} \nu_{\mu}, \tag{4.1}$$

have also been observed. These final states are clearly not CP eigenstates, for they transform into one another under CP. Consequently, they cannot be described in terms of the CP eigenstates K_1^0 and K_2^0; instead they are decay modes of the strangeness eigenstates K^0 and $\bar{K}^0$.

The semileptonic decays of strange particles obey the selection rule

$$\Delta\text{Strangeness} = \Delta\text{Charge} \quad (\Delta S = \Delta Q) \tag{4.2}$$

first postulated by R. Feynman and M. Gell-Mann in 1958. As an example of this rule, the decay

$$\Sigma^- \to n + e^- + \bar{\nu}_e \tag{4.3}$$

is observed (branching ratio $\approx 10^{-3}$), whereas

$$\Sigma^+ \to n + e^+ + \nu_e \tag{4.4}$$

is not (branching ratio $< 5 \times 10^{-6}$).

In the case of neutral kaons, the $\Delta S = \Delta Q$ rule implies

$$K^0 \to e^+ \pi^- \nu_e, \quad \bar{K}^0 \to e^- \pi^+ \bar{\nu}_e. \tag{4.5}$$

A test of (4.2) is shown in Fig. 4.1, where the measured distribution of $K^0 \to e^+ \pi^- \nu_e$ and $\bar{K}^0 \to e^- \pi^+ \bar{\nu}_e$ events from an initially pure K^0 state is plotted as a function of the K^0 decay time [22]. The result is in good agreement with the strangeness oscillation plot of Fig. 1.3, thus confirming the $\Delta S = \Delta Q$ rule (to about 2%).

A related measurement is that of J. Steinberger and his collaborators [23], who obtained the time dependence of the *charge asymmetry*:

$$\mathcal{A}(t) = \frac{N(K^0 \to e^+ \pi^- \nu_e) - N(\bar{K}^0 \to e^- \pi^+ \bar{\nu}_e)}{N(K^0 \to e^+ \pi^- \nu_e) + N(\bar{K}^0 \to e^- \pi^+ \bar{\nu}_e)} \equiv \frac{N^+ - N^-}{N^+ + N^-} \tag{4.6}$$

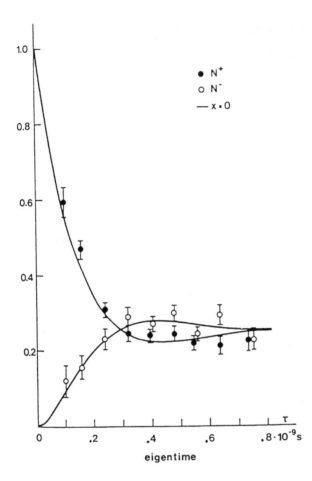

Fig. 4.1. A test of the $\Delta S = \Delta Q$ rule using $K^0 \to e^+\pi^-\nu_e$ and $\bar{K}^0 \to e^-\pi^+\bar{\nu}_e$ decays [22])

(see Fig. 4.2). From expressions for the K^0 and $\bar{K}^0$ probabilities (1.23) and (1.24), it follows[23] that

$$\mathcal{A}(t) = \frac{2e^{-(\Gamma_\mathrm{S}+\Gamma_\mathrm{L})t/2}\cos(\delta_\mathrm{m}t)}{e^{-\Gamma_\mathrm{S}t} + e^{-\Gamma_\mathrm{L}t}} = \frac{2\cos(\delta_\mathrm{m}t)}{e^{-\Delta\Gamma t} + e^{+\Delta\Gamma t}} \tag{4.7}$$

for a pure K^0-state at $t = 0$ ($\Delta\Gamma \equiv (\Gamma_\mathrm{S} - \Gamma_\mathrm{L})/2$ and t is the proper time). Again, the measurement is in good agreement with (4.7) in the strangeness oscillation region.

Note the apparent decay rate asymmetry between the $K^0 \to e^+\pi^-\nu_e$ and $\bar{K}^0 \to e^-\pi^+\bar{\nu}_e$ decays for large values of the K^0 decay time in Fig. 4.2. Assuming the validity of the $\Delta S = \Delta Q$ rule, this asymmetry represents the CP-violating effect mentioned in Sect. 3.1.

---

[23] For a beam of neutral kaons, the probabilities can be replaced by the number densities of particles and antiparticles in the beam.

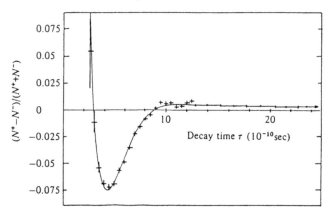

Fig. 4.2. Charge asymmetry as a function of the K_{l3} decay time [23]

In order to study the semileptonic decays $K_{\ell 3} \equiv K^0 \to \ell^\pm \pi^\mp \nu$ in more detail, we define the following transition amplitudes

$$
\begin{aligned}
f &\equiv \langle \ell^+ \pi^- \nu \mid \hat{H}_{\mathrm{w}} \mid K^0 \rangle, & \Delta S &= \Delta Q, \\
g &\equiv \langle \ell^+ \pi^- \nu \mid \hat{H}_{\mathrm{w}} \mid \bar{K}^0 \rangle, & \Delta S &= -\Delta Q.
\end{aligned} \tag{4.8}
$$

Assuming CPT invariance, we have

$$
\begin{aligned}
f^* &\equiv \langle \ell^- \pi^+ \bar{\nu} \mid \hat{H}_{\mathrm{w}} \mid \bar{K}^0 \rangle, & \Delta S &= \Delta Q, \\
g^* &\equiv \langle \ell^- \pi^+ \bar{\nu} \mid \hat{H}_{\mathrm{w}} \mid K^0 \rangle, & \Delta S &= -\Delta Q.
\end{aligned} \tag{4.9}
$$

The operation of $\hat{C}\hat{P}$ transforms particles into antiparticles. If CP is conserved, both f and g are real. The amplitudes g and g^* violate the $\Delta S = \Delta Q$ rule. This violation is small ($g/f \leq 10^{-2}$), as mentioned above. We therefore adjust the relative phase between K^0 and $\bar{K}^0$ so that the $\Delta S = \Delta Q$ amplitudes are real: $f = f^*$ (see footnote 5).

The time development of an initially pure K^0 ($\bar{K}^0$) state can be written as (see (3.17) and (3.32))

$$
|\Psi(t)\rangle = \frac{1}{\sqrt{2}} \left\{ (1 \mp \epsilon) e^{-i\mathcal{M}_{\mathrm{S}} t} |K_{\mathrm{S}}^0\rangle \pm (1 \mp \epsilon) e^{-i\mathcal{M}_{\mathrm{L}} t} |K_{\mathrm{L}}^0\rangle \right\}, \tag{4.10}
$$

where the upper (lower) signs refer to K^0 ($\bar{K}^0$). Expressing K_{S}^0 and K_{L}^0 in terms of K^0 and $\bar{K}^0$ (see (3.16)), we obtain

$$
\begin{aligned}
|\Psi(t)\rangle = \frac{1}{2} \Big\{ & e^{-i\mathcal{M}_{\mathrm{S}} t} \left[|K^0\rangle + (1 - 2\epsilon)|\bar{K}^0\rangle \right] \\
& + e^{-i\mathcal{M}_{\mathrm{L}} t} \left[|K^0\rangle - (1 - 2\epsilon)|\bar{K}^0\rangle \right] \Big\}, \\
\\
|\bar{\Psi}(t)\rangle = \frac{1}{2} \Big\{ & e^{-i\mathcal{M}_{\mathrm{S}} t} \left[|\bar{K}^0\rangle + (1 + 2\epsilon)|K^0\rangle \right] \\
& + e^{-i\mathcal{M}_{\mathrm{L}} t} \left[|\bar{K}^0\rangle - (1 + 2\epsilon)|K^0\rangle \right] \Big\}.
\end{aligned} \tag{4.11}
$$

The transition amplitudes for $\Psi(\overline{\Psi}) \to \ell^{\pm} \pi^{\mp} \nu$ read

$$A_+ = \frac{f}{2} \left[e^{-i\mathcal{M}_{\mathrm{S}}t}(1 + x - 2\epsilon x) + e^{-i\mathcal{M}_{\mathrm{L}}t}(1 - x + 2\epsilon x) \right],$$

$$A_- = \frac{f}{2} \left[e^{-i\mathcal{M}_{\mathrm{S}}t}(x^* + 1 - 2\epsilon) + e^{-i\mathcal{M}_{\mathrm{L}}t}(x^* - 1 + 2\epsilon) \right],$$

$$\overline{A}_+ = \frac{f}{2} \left[e^{-i\mathcal{M}_{\mathrm{S}}t}(x + 1 + 2\epsilon) + e^{-i\mathcal{M}_{\mathrm{L}}t}(x - 1 - 2\epsilon) \right], \quad (4.12)$$

$$\overline{A}_- = \frac{f}{2} \left[e^{-i\mathcal{M}_{\mathrm{S}}t}(1 + x^* + 2\epsilon x^*) + e^{-i\mathcal{M}_{\mathrm{L}}t}(1 - x^* - 2\epsilon x^*) \right],$$

where we defined $x \equiv g/f$. A nonzero value of x implies a violation of the $\Delta S = \Delta Q$ rule; Im $x \neq 0$ indicates CP violation. Squaring the above amplitudes gives the corresponding decay rates (we omit the factor $|f|^2/4$):

$$\Gamma_+ \left[\overline{\Gamma}_- \right] \propto \left\{ |1 + x|^2 \mp 4|x|^2 \mathrm{Re}\ \epsilon \right\} e^{-\Gamma_{\mathrm{S}}t} + \left\{ |1 - x|^2 \mp 4|x|^2 \mathrm{Re}\ \epsilon \right\} e^{-\Gamma_{\mathrm{L}}t}$$
$$+ 2(1 - |x|^2)e^{-\Gamma t} \cos(\delta_{\mathrm{m}}t) \mp 4\mathrm{Im}\ x\, e^{-\Gamma t} \sin(\delta_{\mathrm{m}}t),$$
$$\hspace{10cm} (4.13)$$
$$\Gamma_- \left[\overline{\Gamma}_+ \right] \propto \left\{ |1 + x|^2 \mp 4\mathrm{Re}\ \epsilon \right\} e^{-\Gamma_{\mathrm{S}}t} + \left\{ |1 - x|^2 \mp 4\mathrm{Re}\ \epsilon \right\} e^{-\Gamma_{\mathrm{L}}t}$$
$$- 2 \left\{ (1 - |x|^2) \mp 4\mathrm{Re}\ \epsilon \right\}$$
$$\times e^{-\Gamma t} \cos(\delta_{\mathrm{m}}t) \mp 4\mathrm{Im}\ x\, e^{-\Gamma t} \sin(\delta_{\mathrm{m}}t),$$

where the upper (lower) signs refer to K^0 ($\overline{K}^0$) and $\Gamma \equiv (\Gamma_{\mathrm{S}} + \Gamma_{\mathrm{L}})/2$. Note that the above decay rates depend on the real part of ϵ. Neglecting the small CP violation,

$$\Gamma_{\pm}(\epsilon = 0) \propto |1 + x|^2 e^{-\Gamma_{\mathrm{S}}t} + |1 - x|^2 e^{-\Gamma_{\mathrm{L}}t} \pm 2(1 - |x|^2)e^{-\Gamma t} \cos(\delta_{\mathrm{m}}t)$$
$$- 4\mathrm{Im}\ x\, e^{-\Gamma t} \sin(\delta_{\mathrm{m}}t). \quad (4.14)$$

In the experiment by Niebergall et al. [22], the initial K^0 state was produced in the inelastic *charge-exchange* reaction

$$K^+ p \to K^0 p \pi^+ \quad (4.15)$$

using a 2.4 GeV/c K^+ beam. From a sample of 4724 $K^0 \to e^{\pm} \pi^{\mp} \nu_e$ decays, containing less than 8 background events, they obtained

$$\delta_{\mathrm{m}} \equiv \Delta m_{\mathrm{L,S}} = (0.53 \pm 0.04) \times 10^{10}\,\mathrm{s}^{-1} \quad (4.16)$$

by comparing their data with the theoretical prediction for the time dependence of charge asymmetry in $K_{\ell 3}$ decays:

$$\Gamma_+ - \Gamma_- \propto \left(1 - |x|^2\right) e^{-\Gamma t} \cos(\delta_{\mathrm{m}}t). \quad (4.17)$$

The time distribution of events, together with the theoretical expectation for $x = 0$, is shown in Fig. 4.1. The best fit to the data yielded Re $x = 0.04 \pm 0.03$ and Im $x = -0.06 \pm 0.05$. Combining this result with other measurements referenced in their paper, they found

$$\mathrm{Re}\ x = 0.023 \pm 0.02, \quad \mathrm{Im}\ x = -0.0015 \pm 0.025, \quad (4.18)$$

which is consistent with $x = 0$, i.e., they saw no evidence for the $\Delta S = -\Delta Q$ transitions in semileptonic decays of the neutral kaon.

To determine the time dependence of charge asymmetry

$$\mathcal{A}(t) = \frac{\Gamma_+(t) - \Gamma_-(t)}{\Gamma_+(t) + \Gamma_-(t)}, \quad \bar{\mathcal{A}}(t) = \frac{\overline{\Gamma}_+(t) - \overline{\Gamma}_-(t)}{\overline{\Gamma}_+(t) + \overline{\Gamma}_-(t)} \tag{4.19}$$

($\Gamma_+(t)$ and $\Gamma_-(t)$ are the decay rates to positive and negative leptons, respectively) for an initially pure $K^0(\bar{K}^0)$ beam, consider neutral kaon decays in vacuum close to the beam production point. From (4.13) it follows that

$$\mathcal{A}(t)\left[\bar{\mathcal{A}}(t)\right] = \tag{4.20}$$

$$\frac{2(1 - |x|^2)\left\{\mathrm{Re}\,\epsilon\left[\mathrm{e}^{-\Gamma_{\mathrm{S}}t} + \mathrm{e}^{-\Gamma_{\mathrm{L}}t}\right] \pm \mathrm{e}^{-\Gamma t}\cos(\delta_{\mathrm{m}}t)\right\}}{|1 + x|^2\mathrm{e}^{-\Gamma_{\mathrm{S}}t} + |1 - x|^2\mathrm{e}^{-\Gamma_{\mathrm{L}}t} \pm 4\mathrm{e}^{-\Gamma t}\left[\mathrm{Re}\,\epsilon\cos(\delta_{\mathrm{m}}t) - \mathrm{Im}\,x\sin(\delta_{\mathrm{m}}t)\right]},$$

where the upper (lower) signs refer to K^0 ($\bar{K}^0$). For $x, \epsilon \to 0$ this expression becomes (4.7), as it should.

A neutral kaon beam produced by protons hitting a stationary target is a mixture of K^0 and $\bar{K}^0$ mesons. In fact, it is an incoherent mixture because K^0 and its antiparticle are produced differently. Consequently, the interference terms in (4.20) have to be multiplied by the *dilution factor* $\mathcal{D}(p)$: $\pm\cos(\delta_{\mathrm{m}}t) \to \mathcal{D}(p)\cos(\delta_{\mathrm{m}}t)$, $\mp\sin(\delta_{\mathrm{m}}t) \to -\mathcal{D}(p)\sin(\delta_{\mathrm{m}}t)$ (see (4.42)); $\mathcal{D}(p)$ is defined by

$$\mathcal{D}(p) \equiv \frac{S(p) - \bar{S}(p)}{S(p) + \bar{S}(p)}, \tag{4.21}$$

where $S(p)$ and $\bar{S}(p)$ are momentum-dependent production intensities of K^0 and $\bar{K}^0$ mesons, respectively. Note that for $S(p) = \bar{S}(p)$ there would be no interference at all. However, because of the lower production threshold for K^0 (Sect. 1.5), $S(p) \gg \bar{S}(p)$ in general. In the interference region $\mathrm{e}^{-\Gamma_{\mathrm{L}}t} \approx 1$ and $\mathrm{e}^{-\Gamma_{\mathrm{S}}t} \ll 1$, and so (4.20) simplifies to

$$\mathcal{A}(t) = \frac{2(1 - |x|^2)}{|1 - x|^2}\left\{\mathcal{D}(p)\mathrm{e}^{-(\Gamma_{\mathrm{S}} - \Gamma_{\mathrm{L}})t/2}\cos(\delta_{\mathrm{m}}t) + \mathrm{Re}\,\epsilon\right\}. \tag{4.22}$$

The apparatus employed by J. Steinberger and his coworkers in the experiment mentioned above is described in Sect. 4.3. Semileptonic decays were selected from the data by an unambiguous lepton signature on one of the two charged decay products of K^0. The reconstructed momenta of charged tracks were restricted to (a) $p_e, p_\pi < 8\,\mathrm{GeV}/c$ in order to eliminate pions above the Čerenkov threshold (8.4 GeV/c) from the K_{e3} sample, and (b) $p_\mu > 1.6\,\mathrm{GeV}/c$, since the minimum momentum to reach muon counters was 1.45 GeV/c. In total, 6 million K_{e3} and 2 million $K_{\mu3}$ decays with lifetimes $t \leq 12.75 \times 10^{-10}s$ remained after these and other cuts to remove possible backround events.

The mass difference $\Delta m_{\mathrm{L,S}}$ was determined from a comparison of the measured $\mathcal{A}(t)$ distribution (see Fig. 4.2) with the theoretical prediction (4.20),

(4.22), taking into account radiative corrections, experimental resolution and acceptance, etc., with the result

$$\Delta m_{\mathrm{L,S}} = (0.5334 \pm 0.004) \times 10^{10}\,\mathrm{s}^{-1}. \tag{4.23}$$

From (4.22) we see that

$$\mathcal{A}(t) \longrightarrow \begin{cases} \gg 2\mathrm{Re}\,\epsilon, & \Gamma_{\mathrm{S}}t \approx 1, \\ \approx 2\mathrm{Re}\,\epsilon, & \Gamma_{\mathrm{S}}t \gg 1. \end{cases} \tag{4.24}$$

Therefore, at $\Gamma_{\mathrm{S}}t \gg 1$ the charge asymmetry is given by the second term in (4.22):

$$\mathcal{A}_{\mathrm{L}}(t) \equiv \frac{\Gamma(K_{\mathrm{L}}^0 \to \ell^+\pi^-\nu) - \Gamma(K_{\mathrm{L}}^0 \to \ell^-\pi^+\nu)}{\Gamma(K_{\mathrm{L}}^0 \to \ell^+\pi^-\nu) + \Gamma(K_{\mathrm{L}}^0 \to \ell^-\pi^+\nu)} \approx 2\mathrm{Re}\,\epsilon\,\frac{1 - |x|^2}{|1 - x|^2}. \tag{4.25}$$

Assuming that $x = 0$, a measurement of $\mathcal{A}_{\mathrm{L}}(t)$ at $\Gamma_{\mathrm{S}}t \gg 1$ yields $2\mathrm{Re}\,\epsilon = \langle K_{\mathrm{L}}^0 \mid K_{\mathrm{S}}^0 \rangle$.

To keep the effects of K_{S}^0-K_{L}^0 interference as low as possible, they selected events according to $t_{e3} > 12.75 \times 10^{-10}\,\mathrm{s}$ and $t_{\mu3} > 14.75 \times 10^{-10}\,\mathrm{s}$. Based on a total of 34 million K_{e3} and 15 million $K_{\mu3}$ events, their measurement yielded (see [24])

$$\mathrm{Re}\,\epsilon = (1.67 \pm 0.08) \times 10^{-3}. \tag{4.26}$$

Using (3.80)

$$\mathrm{Re}\,\epsilon = \frac{|\eta_{+-}|}{\sqrt{1 + (2\Delta m_{\mathrm{L,S}}/\Gamma_{\mathrm{S}})^2}} \tag{4.27}$$

and the measured values of $\Delta m_{\mathrm{L,S}}$, $|\eta_{+-}|$ and Γ_{S} (see Sect. 4.3), they computed

$$\frac{|\eta_{+-}|}{\sqrt{1 + (2\Delta m_{\mathrm{L,S}}/\Gamma_{\mathrm{S}})^2}} = (1.66 \pm 0.03) \times 10^{-3}, \tag{4.28}$$

in good agreement with the above result.

4.2 K_{L}^0-K_{S}^0 Interference in $\pi^+\pi^-$ and $\ell^\pm\pi^\mp\nu$ Decays

We now describe a high-precision K_{L}^0-K_{S}^0 interference experiment which measured the phase difference $\phi_{+-} - \phi_\varrho$ using the pionic decay modes $K_{\mathrm{L,S}}^0 \to \pi^+\pi^-$ and, concurrently, the regeneration phase ϕ_ϱ from the time-dependent charge asymmetry in $K_{\mathrm{L,S}}^0 \to e^\pm\pi^\mp\nu_e$ and $\mu^\pm\pi^\mp\nu_\mu$ decays [25].

In this experiment, a K_{L}^0 beam of 4 to 10 GeV/c momentum from the Brookhaven AGS traversed an 81-cm-long block of carbon. Multiwire proportional chambers were used to measure time distributions of $\pi^+\pi^-$ and semileptonic decays behind the regenerator.

The semileptonic decay rates Γ_+ and Γ_- behind a regenerator are obtained from (3.89) by repeating the steps which led to (4.13), with the result

$$\Gamma_\pm(\varrho_c) \propto \pm 2\mathrm{Re}\,\epsilon(1-|x|^2)\left\{|\varrho_c|^2\mathrm{e}^{-\Gamma_S t} + \mathrm{e}^{-\Gamma_L t}\right\}$$
$$+\,|1+x|^2|\varrho_c|^2\mathrm{e}^{-\Gamma_S t} + |1-x|^2\mathrm{e}^{-\Gamma_L t}$$
$$+\,2\left\{2\mathrm{Re}\,\epsilon \pm (1-|x|^2)\right\}|\varrho_c|$$
$$\times\,\mathrm{e}^{-\Gamma t}\cos(\delta_m t + \phi_\varrho) - 4|\varrho_c|\mathrm{Im}\,x\,\mathrm{e}^{-\Gamma t}\sin(\delta_m t + \phi_\varrho). \quad (4.29)$$

In the interference region, $\mathrm{e}^{-\Gamma_L t} \approx 1$ and $\mathrm{e}^{-\Gamma_S t} \ll 1$. Observing also that $|\varrho_c|^2 \ll 1$, (4.29) yields

$$\mathcal{A}(t) \approx \frac{2(1-|x|^2)}{|1-x|^2}\left\{|\varrho_c|\mathrm{e}^{-(\Gamma_S-\Gamma_L)t/2}\cos(\delta_m t + \phi_\varrho) + \mathrm{Re}\,\epsilon\right\}, \quad (4.30)$$

which is the expression used by W. Carithers et al. [25]

The measured charge asymmetry of $K_{e3}^0 \equiv K^0 \to e^\pm\pi^\mp\nu_e$ and $K_{\mu 3}^0 \equiv K^0 \to \mu^\pm\pi^\mp\nu_\mu$ decays as a function of proper time is shown in Fig. 4.3. From this asymmetry they extracted the nuclear regeneration phase, ϕ_f, defined by

$$\phi_\varrho = \phi_f + \phi_\zeta, \quad (4.31)$$

where

$$\phi_f \equiv \arg\left[\mathrm{i}(\mathsf{f}-\bar{\mathsf{f}})\right], \quad \phi_\zeta \equiv \arg\left[\frac{1-\mathrm{e}^{-(-\mathrm{i}\delta_m+1/2)\zeta}}{-\mathrm{i}\delta_m + 1/2}\right]. \quad (4.32)$$

Namely, they split the phase of the coherent regeneration amplitude (see (2.39))

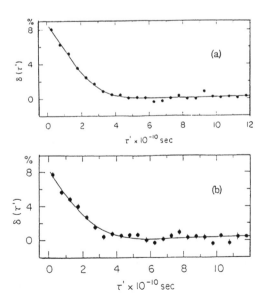

Fig. 4.3. (a) K_{e3} charge asymmetry versus proper time; (b) $K_{\mu 3}$ charge asymmetry [25]

$$\varrho_{\mathrm{c}} \equiv |\varrho_{\mathrm{c}}|e^{i\phi_\varrho} = \frac{N\lambda_{\mathrm{S}}\varLambda_{\mathrm{S}}}{2}\,[i(\mathfrak{f}-\bar{\mathfrak{f}})]\,\frac{1-e^{-(-i\delta_{\mathrm{m}}+1/2)\zeta}}{-i\delta_{\mathrm{m}}+1/2} \tag{4.33}$$

into two parts, one of which, ϕ_ζ, can be easily computed.

The data were simultaneously fitted for $\chi|\varrho_{\mathrm{c}}|$ and ϕ_ϱ, with $\Delta m_{\mathrm{L,S}}$, $\varGamma_{\mathrm{S}}$ and $\mathcal{A}_{\mathrm{L}}$ fixed; χ and $\mathcal{A}_{\mathrm{L}}$ are defined as (see (4.25))

$$\chi \equiv \frac{2(1-|x|^2)}{|1-x|^2}, \quad \mathcal{A}_{\mathrm{L}}(t) \equiv 2\chi\operatorname{Re}\epsilon = \chi\langle K^0_{\mathrm{L}} \mid K^0_{\mathrm{S}}\rangle. \tag{4.34}$$

$\mathcal{A}_{\mathrm{L}}(t)$ is the charge asymmetry in the decays $K^0_{\mathrm{L}} \to \ell^\pm\pi^\mp\nu$, first observed by J. Steinberger and his colleagues at Brookhaven [26], and by D. Dorfan et al. at SLAC [27]. The value of $\mathcal{A}_{\mathrm{L}}(t)$ used by Carithers et al. is from [24]. The combined K_{e3} and $K_{\mu3}$ data yielded

$$\phi_f = -41^0 \pm 2.6^\circ. \tag{4.35}$$

The $\pi^+\pi^-$ intensity in the forward direction behind a regenerator placed in a K^0_{L}-beam is given by (3.91). Similar to the first part of their analysis, the $\pi^+\pi^-$ data (see Fig. 4.4) were fitted simultaneously for $|\varrho_{\mathrm{c}}/\eta_{+-}|$, $\phi_{+-}-\phi_f$ and $\varGamma_{\mathrm{S}}$, with $\Delta m_{\mathrm{L,S}}$ and $\varGamma_{\mathrm{L}}$ fixed. Combining the result for $\phi_{+-}-\phi_f$ with the measurement of ϕ_f (see Fig. 4.5), the CP violating phase η_{+-} was found to be

$$\phi_{+-} = 45.5^\circ \pm 2.8^\circ. \tag{4.36}$$

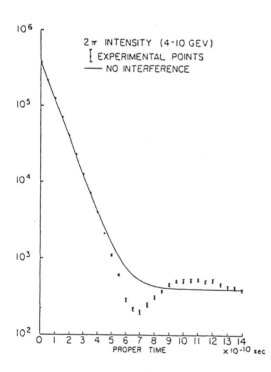

Fig. 4.4. Rate of $\pi^+\pi^-$ events as a function of proper time downstream from a carbon regenerator (T. Modis, Columbia Univ. thesis, 1973; and [25])

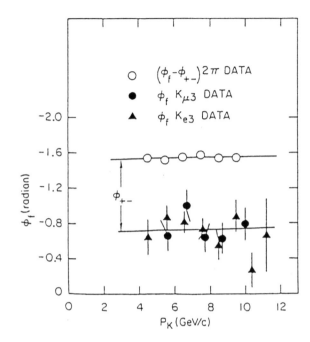

Fig. 4.5. Results of the combined fit for ϕ_{+-} [25]

Their analysis can be summarized as follows:

$$\left.\begin{array}{l} \ell^\pm \pi^\mp \nu, \quad [\delta_m t + (\phi_f + \phi_\zeta)] \longrightarrow \phi_f \approx -41° \\ \pi^+ \pi^-, \quad [\delta_m t + (\phi_f + \phi_\zeta) - \phi_{+-}] \longrightarrow \phi_f - \phi_{+-} \approx -87° \end{array}\right\},$$

$$\phi_{+-} \approx 46°. \tag{4.37}$$

4.3 K_L^0-K_S^0 Interference Without Regenerator

The CP violating phase ϕ_{+-} can be measured independently of the regeneration phase ϕ_ϱ by observing K_L^0-K_S^0 interference in vacuum close to the $K^0(\bar{K}^0)$ production point. This so-called *vacuum interference* method requires the mass difference $\Delta m_{L,S}$ to be known very accurately: a 1% error in $\Delta m_{L,S}$ corresponds to an uncertainty of 3° in ϕ_{+-}.

Here we describe an experiment by J. Steinberger and his collaborators at the CERN Proton Synchrotron,[24] who used results from two high precision measurements of $\Delta m_{L,S}$ in the same detector [23, 28] to obtain ϕ_{+-}. They were the first to employ multiwire proportional chambers (MWPC), invented by G. Charpak. MWPCs can handle event rates that are hundreds of times higher than those possible with spark chambers. The detector, described below, was also used in the charge asymmetry measurement discussed earlier.

---

[24] A CERN–Heidelberg collaboration.

The apparatus is sketched in Fig. 4.6. The neutral kaons, produced by 24 GeV/c protons hitting a platinum target, were selected by a collimator at an angle of roughly 75 mrad. Protons with such momenta produce at small angles about three times as many K^0s as $\bar{K}^0$s. The kaon momenta were in the range 3–15 GeV/c.

The collimator was followed by a 9-m-long decay volume filled with helium ("cheap vacuum"). A 6-m-long threshold Čerenkov counter, containing hydrogen gas at atmospheric pressure, was used to identify electrons. Muons were identified by two counters behind a concrete absorber at the far end of the detector. The decay region, extending 2.2 m to 11.6 m after the target, permitted detection in the proper time interval 3.5×10^{-10} s $< t_{\mathrm{p}} < 30 \times 10^{-10}$ s. The momenta of charged decay products of neutral kaons were measured in a spectrometer consisiting of four MWPCs and a bending magnet. A total of 10^9 events was registered, with an average rate of about 1000 events per machine cycle.

To select $K^0_{\mathrm{L,S}} \to \pi^+\pi^-$ events only inward bending pairs of charged particles were retained. They also required that: (a) there must be no signal in the Čerenkov counter and no coincidence between the two muon counters; and (b) the momenta of both particles must lie in the interval 1.5 GeV/c to 8.5 GeV/c, i.e., above the minimum momentum to traverse the muon absorber (1.45 GeV/c) and below the threshold for pion detection in the Čerenkov counter (8.4 GeV/c).

To derive the $\pi\pi$ decay distribution from the interfering K^0_{L} and K^0_{S} states, let us assume that at $t = 0$ a pure $K^0(\bar{K}^0)$ beam is produced. At a later time t (see (4.10))

$$|\Psi(t)\rangle = \frac{1}{\sqrt{2}} \left\{ (1 \mp \epsilon)\mathrm{e}^{-\mathrm{i}\mathcal{M}_S t}|K^0_{\mathrm{S}}\rangle \pm (1 \mp \epsilon)\mathrm{e}^{-\mathrm{i}\mathcal{M}_L t}|K^0_{\mathrm{L}}\rangle \right\}, \qquad (4.38)$$

where the upper (lower) signs refer to a K^0 ($\bar{K}^0$). Note that the phase between K^0_{L} and K^0_{S} at $t = 0$ is $0°$ ($180°$) if the original state is a K^0 ($\bar{K}^0$).

The decay amplitudes read

$$\mathsf{A}_{\pi\pi} = \mathsf{A}_{K^0_{\mathrm{S}} \to \pi\pi} \frac{1}{\sqrt{2}} \mathrm{e}^{-\mathrm{i}m_S t}$$

$$\times \left\{ (1 \mp \epsilon)\mathrm{e}^{-\Gamma_S t/2} \pm (1 \mp \epsilon)|\eta|\mathrm{e}^{-\mathrm{i}(\delta_m t - \phi) - \Gamma_L t/2} \right\} \qquad (4.39)$$

and the corresponding decay rates (we omit the factor $\Gamma_{S,+-}/2$, i.e., $\Gamma_{S,00}/2$)

$$\Gamma_{\pi\pi} \propto (1 \mp 2\mathrm{Re}\,\epsilon)$$

$$\times \left\{ |\eta|^2 \mathrm{e}^{-\Gamma_L t} + \mathrm{e}^{-\Gamma_S t} \pm 2|\eta|\mathrm{e}^{-(\Gamma_S + \Gamma_L)t/2} \cos(\delta_m t - \phi) \right\}. \qquad (4.40)$$

The interference term changes sign when K^0 is replaced by $\bar{K}^0$, resulting in different $\pi\pi$ decay distributions for the two states, as shown in Fig. 4.7. This illustrates nicely the violation of *CP* symmetry.

The distribution (4.40) is practically identical to that behind a regenerator (see (3.91)), with the regeneration amplitude $\varrho_c = 1$.

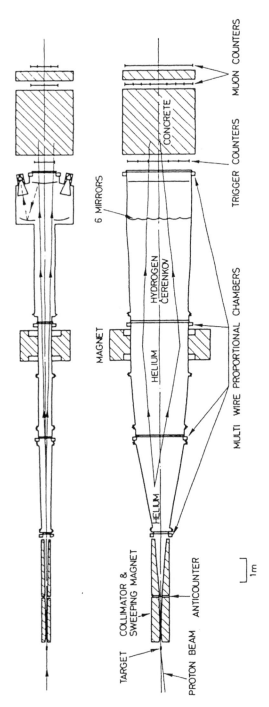

Fig. 4.6. The apparatus used by J. Steinberger and his collaborators at CERN

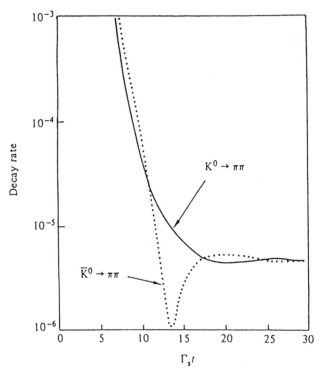

Fig. 4.7. Difference in $\pi\pi$ decay distributions between initially pure K^0 and $\bar{K}^0$ beams, illustrating the violation of CP symmetry

The magnitude of η_{+-} was obtained by measuring the $\pi^+\pi^-$ decay rate at $\Gamma_S t \gg 1$, $\Gamma_L t \ll 1$ ($t \geq 15 \times 10^{-10}$ s). The phase of η_{+-} was measured in the interference region (5×10^{-10} s $\leq t \leq 15 \times 10^{-10}$ s) by isolating the cosine-term. The results were [29, 30]

$$\phi_{+-} = 45.9° \pm 1.6°, \quad |\eta_{+-}| = (2.3 \pm 0.035) \times 10^{-3}. \tag{4.41}$$

To obtain ϕ_{+-} they used $\Delta m_{\mathrm{L,S}} = (0.5338 \pm 0.00215) \times 10^{10}$ s^{-1}, the combined value of the two measurements of $\Delta m_{\mathrm{L,S}}$ in the same detector referenced earlier (one measurement was based on the *variable gap* regeneration method, and the other one on charge asymmetry in semileptonic K^0 decays).

In fact, their analysis was slightly more complicated than this simplified description because the neutral kaon beam is an incoherent mixture of K^0 and $\bar{K}^0$ particles, as explained in Sect. 4.1. The $\pi\pi$ decay intensity is therefore a linear combination of the two distributions (4.40):

$$\Gamma_{\pi\pi} \propto \left[S(p) + \bar{S}(p) \right]$$
$$\times \left\{ |\eta|^2 e^{-\Gamma_L t} + e^{-\Gamma_S t} + 2\mathcal{D}(p)|\eta| e^{-(\Gamma_S + \Gamma_L)t/2} \cos(\delta_m t - \phi) \right\} \tag{4.42}$$

(see (4.21)). Expression (4.42) was fitted to the $\pi^+\pi^-$ data to obtain Γ_S, $|\eta_{+-}|, \phi_{+-}, S(p)$ and $\bar{S}(p)$, assuming that $\Delta m_{L,S}$ and Γ_L are known (see Fig. 4.8). The fit yielded

$$\Gamma_S = (1.119 \pm 0.006) \times 10^{10}\,\text{s}^{-1}. \tag{4.43}$$

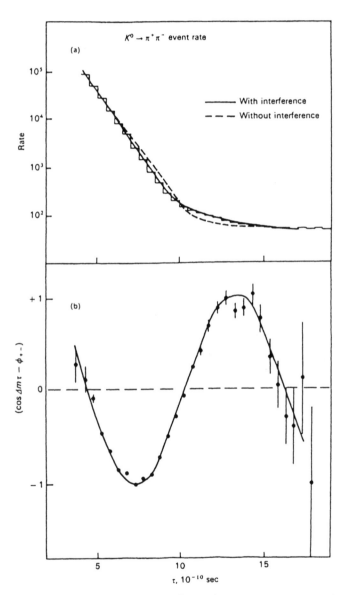

Fig. 4.8. (a) The measured $K^0 \to \pi^+\pi^-$ event rate and (b) the extracted interference term [29]

5. Precision Measurements of ϕ_{00}, ϕ_{+-} and ε'/ε

The decay modes $K^0_{L,S} \to \pi^0\pi^0$ are considerably more difficult to investigate experimentally than their charge counterparts $K^0_{L,S} \to \pi^+\pi^-$. The reason is that neutral pions cannot be observed directly: a π^0 decays within 10^{-16} s into two photons. Instead, one has to detect and measure electromagnetic "showers" associated with the reaction shown in Fig. 5.1. The difficulty lies in measuring accurately the direction and energy of the final state photons.

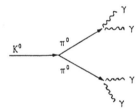

Fig. 5.1. $K^0 \to \pi^0\pi^0 \to 4\gamma$

An additional complication arises from the fact that the CP conserving decay $K^0_L \to 3\pi^0 \to 6\gamma$ is two hundred times more frequent than the CP violating decay $K^0_L \to 2\pi^0 \to 4\gamma$. Since $3\pi^0$ decays can simulate 4γ events if two photons are not detected, one has to rely on distinct kinematical features of the $2\pi^0$ decay mode to eliminate this background.

As the photon energy increases, its measurement precision improves. The two experiments described below used intense fluxes of neutral kaons with energies of about 100 GeV. High beam intensities are essential to achieve adequate statistical accuracy, especially in the $K^0_L \to 2\pi^0$ channel. Both experiments measured the CP-violating parameter ε'/ε, which can be related to the $K^0_{L,S} \to \pi^0\pi^0, \pi^+\pi^-$ decay intensities in the following way.

Using

$$|\eta_{00}|^2 = \frac{\Gamma_{L,00}}{\Gamma_{S,00}}, \quad |\eta_{+-}|^2 = \frac{\Gamma_{L,+-}}{\Gamma_{S,+-}} \tag{5.1}$$

(see (3.46)) we form the double ratio of decay intensities

$$\frac{|\eta_{00}|^2}{|\eta_{+-}|^2} = \frac{\Gamma(K^0_L \to \pi^0\pi^0)/\Gamma(K^0_S \to \pi^0\pi^0)}{\Gamma(K^0_L \to \pi^+\pi^-)/\Gamma(K^0_S \to \pi^+\pi^-)}. \tag{5.2}$$

From expressions (3.72) and (3.73) it follows that

$$\eta_{00} \approx \varepsilon \left(1 - 2\frac{\varepsilon'}{\varepsilon} \right), \quad \eta_{+-} \approx \varepsilon \left(1 + \frac{\varepsilon'}{\varepsilon} \right). \tag{5.3}$$

Hence,

$$| \eta_{00} |^2 \approx |\varepsilon|^2 \left[1 - 4\operatorname{Re}\left(\frac{\varepsilon'}{\varepsilon}\right) \right], \tag{5.4}$$

$$|\eta_{+-}|^2 \approx |\varepsilon|^2 \left[1 + 2\operatorname{Re}\left(\frac{\varepsilon'}{\varepsilon}\right) \right] \tag{5.5}$$

and

$$\operatorname{Re}\left(\frac{\varepsilon'}{\varepsilon}\right) \approx \frac{1}{6} \left[1 - \frac{| \eta_{00} |^2}{|\eta_{+-}|^2} \right]. \tag{5.6}$$

By observing all four $K_{L,S}^0 \to \pi^0\pi^0, \pi^+\pi^-$ decay modes simultaneously, or at least two at a time, beam intensities and detection efficiencies cancel in the double ratio (5.2), thus minimizing systematic uncertainties in the measurement of ε'/ε.

5.1 The Experiment NA31 at CERN

We first describe an experiment by J. Steinberger and his coworkers at the CERN Super Proton Synchrotron (SPS).[25] Intense beams of K_L^0 and K_S^0 mesons with energies around 100 GeV were produced *alternately* by 450 GeV protons (10^{11} and 10^7 protons per pulse, respectively) at two different targets (see Fig. 5.2). The $\pi^0\pi^0$ and $\pi^+\pi^-$ decay modes were detected *concurrently*, however. The K_S^0 data were taken with the corresponding target displaced in steps of 1.2 m, which resulted in uniform K_S^0 and K_L^0 decay distributions over a 48-m-long decay region, despite the short K_S^0 decay length (6 m on the average).

The decay region was evacuated and the space between two tracking wire chambers, set 25 m apart, was filled with helium. Photons from π^0 decays were measured in a liquid-argon/lead calorimeter which was also used, together with an iron/scintillator calorimeter, to measure the energy of charged pions. There was no magnetic spectrometer.

The $K_{L,S}^0 \to \pi^+\pi^-$ decays were reconstructed from hits in the two wire chambers, and the $K_{L,S}^0 \to \pi^0\pi^0$ decays from the measured positions and energies of the photons. The energy spectra of accepted $\pi^0\pi^0$ and $\pi^+\pi^-$ events are shown in Fig. 5.3. After corrections for various systematic uncertainties, the analysis, based on their 1986 data, yielded

$$\operatorname{Re}\left(\frac{\varepsilon'}{\varepsilon}\right) = (3.3 \pm 1.1) \times 10^{-3}, \tag{5.7}$$

---

[25] The NA31 Collaboration.

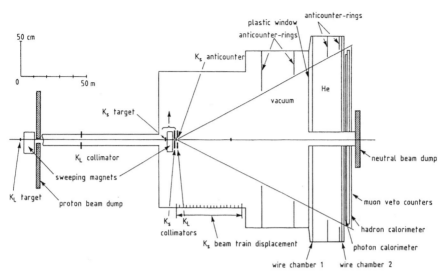

Fig. 5.2. The NA31 experimental set-up at CERN

which was interpreted as the first evidence of direct CP violation in the $K_L^0 \to 2\pi$ decay [31a]. Using data collected in 1988 and 1989, they reported $\mathrm{Re}\,(\varepsilon'/\varepsilon) = (2.0 \pm 0.7) \times 10^{-3}$ [31b].

The NA31 collaboration also determined the phases of the CP-violating parameters η_{00} and η_{+-} from the time dependence of the 2π decay rates by using (4.42). $\Gamma_{\pi\pi}$ is most sensitive to ϕ in the region where K_S^0 and K_L^0 decay rates are about equal ($\approx 12 K_S^0$ lifetimes). The original beam layout was therefore modified to obtain the maximum acceptance in the interference region. The phases ϕ_{+-} and ϕ_{00} were determined from the ratio of decay distributions for two different target positions (one near to the detector, $\mathcal{K}_{\mathrm{near}}$, and one far from the detector, $\mathcal{K}_{\mathrm{far}}$), which renders the acceptance correction negligible. The maximum sensitivity is obtained when the interference patterns from the two targets are displaced by $\pi/2$; at 100 GeV/c this corresponds to a distance of about 15 m.

The measured decay rates from the combined $\mathcal{K}_{\mathrm{near}}$ and $\mathcal{K}_{\mathrm{far}}$ data are shown in Fig. 5.4. The interference term was extracted by subtracting the fitted lifetime distribution without interference from the data. The phases ϕ_{+-} and ϕ_{00} were obtained in the following way. They took the ratio of events observed from the two targets (see Fig. 5.5). A simultaneous fit to $(\mathcal{K}_{\mathrm{near}}/\mathcal{K}_{\mathrm{far}})_{00}$ and $(\mathcal{K}_{\mathrm{near}}/\mathcal{K}_{\mathrm{far}})_{+-}$ was made using bins of 5 GeV/c momentum and $0.5\tau_S$ (τ_S was computed for each event from the mid-point between the two targets). The phases ϕ_{+-} and ϕ_{00} and the dilution factor $\mathcal{D}(p)$ were varied in the fit, while $\tau_S, \tau_L, |\eta_{+-}|, |\eta_{00}|$ and $\Delta m_{\mathrm{L,S}}$ were fixed. Taking into account various systematic uncertainties, they obtained [32]

$$
\begin{aligned}
\phi_{+-} &= (46.9 \pm 2.2)^\circ, \\
\phi_{00} &= (47.1 \pm 2.8)^\circ \longrightarrow \phi_{+-} - \phi_{00} = (0.2 \pm 2.9)^\circ.
\end{aligned}
\tag{5.8}
$$

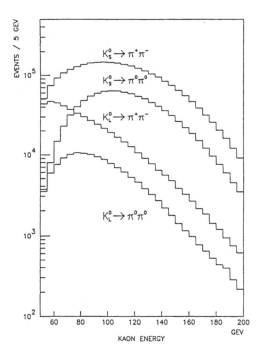

Fig. 5.3. Energy spectra of $\pi^0\pi^0$ and $\pi^+\pi^-$ decays and the corresponding event statistics [31a]

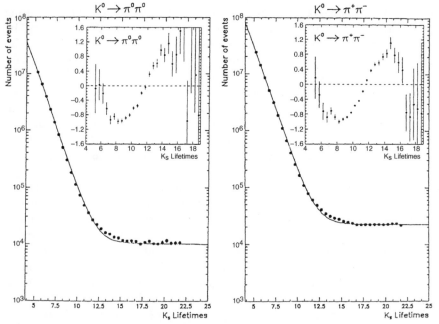

Fig. 5.4. Acceptance-corrected lifetime distributions. Insets show the difference between a fit without the interference term and the data, averaged over energy [32]

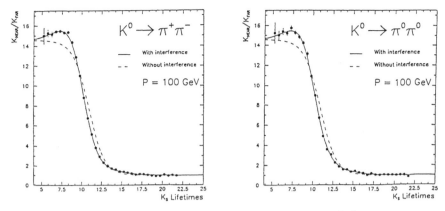

Fig. 5.5. Ratio of decay distributions for two target positions [32]

5.2 The Experiment E731/E773 at Fermilab

The second experiment was performed at Fermilab by B. Winstein and his collaborators,[26] who used 800 GeV protons incident on a berillium target to produce two parallel kaon beams, one pure K_L^0 and one with coherently regenerated K_S^0 mesons (see Fig. 5.6). In this experiment $|\varrho_c| \approx 10 \times |\eta|$, and so the 2π-decays from the regenerator beam were mostly K_S^0 mesons. This way they obtained K_L^0 and K_S^0 beams with almost identical momentum and spatial distributions. The regenerator alternated between the beams once

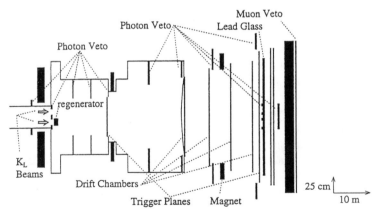

Fig. 5.6. Detector layout of the experiment E731 at Fermilab

every minute, thus essentially eliminating any small difference in beam intensity or detector acceptance for decays from the two beams. However, since $\tau_L \gg \tau_S$, the detector acceptance as a function of decay vertex must be

---

[26] The E731/E773 Collaborations.

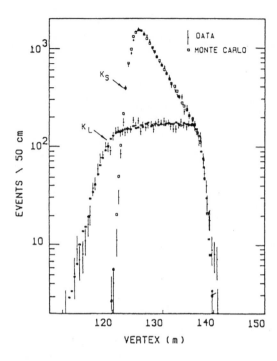

Fig. 5.7. $\pi^0\pi^0$ decay positions [33]

precisely known. This acceptance was determined by using a highly detailed Monte Carlo simulation which relied on $K_{\ell 3}$ and $3\pi^0$ decays. To minimize systematic uncertainties, K_L^0 and K_S^0 decays to $\pi^0\pi^0$ and $\pi^+\pi^-$ final states were detected *simultaneously*.

A drift chamber spectrometer was employed to determine the $\pi^+\pi^-$ momenta, mass and decay vertex. The energies and positions of the four photons from the $\pi^0\pi^0$ decays were measured with a lead-glass calorimeter. The $K_{L,S}^0 \to \pi^0\pi^0$ decay position and the $\pi^0\pi^0$ effective mass were obtained from the best pairing of photons into two pions (see Figs. 5.7 and 5.8 [33]).

Semileptonic events were removed from the $\pi^+\pi^-$ sample using the ratio of shower energy to track momentum (for K_{e3} decays) and a muon "hodoscope" (for $K_{\mu 3}$ decays). The $3\pi^0$ backround to the $\pi^0\pi^0$ data was estimated by Monte Carlo calculations (Fig. 5.8).

After background subtraction, the full E731 data set contained ($3.27 \times 10^5\pi^+\pi^-$, $4.1\times 10^5\pi^0\pi^0$) *vacuum events* and ($1.06\times 10^6\pi^+\pi^-$, $8.0\times 10^5\pi^0\pi^0$) *regenerator events*. The data were collected in 1987 and 1988 at the Fermilab "Tevatron" accelerator.

To obtain Re $(\varepsilon'/\varepsilon)$, the ratio of vacuum to regenerator events was fitted in momentum (p) and decay position (z) bins by using the following expression for the event rate downstream of the regenerator:

$$\frac{\mathrm{d}N}{\mathrm{d}p\mathrm{d}z} \propto \left| \varrho_{\rm c}(p)\mathrm{e}^{-z(\Gamma_{\rm S}/2 - \mathrm{i}\Delta m_{\rm L,S})/\beta\gamma c} + \eta\mathrm{e}^{-z\Gamma_{\rm L}/2\beta\gamma c} \right|^2 , \tag{5.9}$$

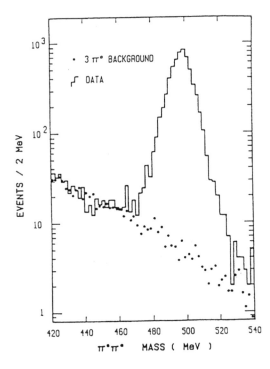

Fig. 5.8. $\pi^0\pi^0$ effective mass [33]

where c is the speed of light in vacuum, $\gamma = E_k/m_k$, $\eta_{+-} = \varepsilon(1 + \varepsilon'/\varepsilon)$ and $\eta_{00} = \varepsilon(1 - 2\varepsilon'/\varepsilon)$. In the fit they used (a) their own values of $\Delta m_{\mathrm{L,S}}$ and Γ_{S}, (b) the world average of $|\eta_{+-}|$ for $|\varepsilon|$, (c) $\phi_\varepsilon = \arctan(2\Delta m_{\mathrm{L,S}}/\Gamma_{\mathrm{S}})$, (d) $\phi_{\varepsilon'} = (43 \pm 6)^0$, and (e) the empirical power-law parametrization of the regeneration amplitude

$$\varrho_c(p) \propto p^{-\alpha}e^{-i(2-\alpha)\pi/2} \tag{5.10}$$

(see [34] regarding the above parametrization).

Fits were first done for each decay mode separately, setting $\varepsilon' = 0$, to extract α and ϱ_c. The $\pi^+\pi^-$ and $\pi^0\pi^0$ results were found to be mutually consistent, which points to a small value of ε'/ε.

A grand fit was next made to both modes simultaneously for the value of $\mathrm{Re}\,(\varepsilon'/\varepsilon)$, allowing the regeneration parameters to vary. They obtained [35]

$$\mathrm{Re}\left(\frac{\varepsilon'}{\varepsilon}\right) = (0.74 \pm 0.59) \times 10^{-3}, \tag{5.11}$$

a value not significantly different from zero.

The full E731 data set was also used to measure the neutral kaon parameters $\Delta m_{\mathrm{L,S}}$, τ_{S}, ϕ_{+-} and $\Delta\phi \equiv \phi_{00} - \phi_{+-}$. To extract $\Delta m_{\mathrm{L,S}}$ and τ_{S} they fixed

$$\phi_{00} \approx \phi_{+-} \approx \arctan\left(\frac{2\Delta m_{\mathrm{L,S}}}{\Gamma_{\mathrm{S}}}\right) = 43.7^\circ \tag{5.12}$$

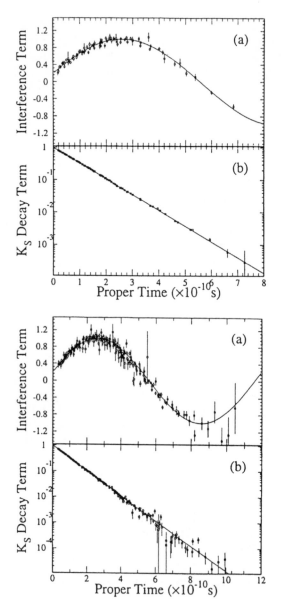

Fig. 5.9a,b. Distributions in proper time for $\pi^+\pi^-$ decays. The lines are the best fit results described in [35]

Fig. 5.10a,b. Distributions in proper time for $2\pi^0$ decays. The lines are the best fit results described in [35]

and simultaneously varied τ_S, $\Delta m_{L,S}$, α and ϱ_c ($p = 70\,\text{GeV/c}$) in expressions (5.9) and (5.10)). The extracted interference and exponential terms, together with the superposed best fits, are shown in Figs. 5.9a,b and 5.10a,b [35].

Combining the values for $\Delta m_{L,S}$ and τ_S obtained from the two decay modes, they found

$$\Delta m_{L,S} = (0.5286 \pm 0.0028) \times 10^{10}\,\text{s}^{-1},$$
$$\tau_S = (0.8929 \pm 0.0016) \times 10^{-10}\,\text{s}. \tag{5.13}$$

The $\Delta m_{\mathrm{L,S}}$ result was lower than the existing world average by about two standard deviations. Based on the reported dependences upon $\Delta m_{\mathrm{L,S}}$, they corrected the best previous measurements of ϕ_{+-} by using their value of $\Delta m_{\mathrm{L,S}}$. The corrected values were found to be in excellent agreement with each other and with (5.12), as expected from *CPT* symmetry.

To extract $\Delta\phi$, a simultaneous fit to the charged and neutral data was made, allowing ϕ_{+-}, $\Delta\phi$ and ε'/ε to vary, with the result[27]

$$\phi_{00} - \phi_{+-} = (-1.6 \pm 1.2)^\circ. \tag{5.14}$$

A similar fit with $\Delta m_{\mathrm{L,S}}$ floating yielded

$$\phi_{+-} = (42.2 \pm 1.4)^\circ \tag{5.15}$$

in agreement with (5.12), which is based on the world-average values for $\Delta m_{\mathrm{L,S}}$ and Γ_{S}.

The apparatus of experiment E773, which took data in 1991, was essentially the same as that of experiment E731, the main difference being that K_{L}^0 mesons this time impinged on two different regenerators, one placed 117 m and the other one 128 m from the target. For this run a new "active" regenerator made of plastic scintillator was used, thereby reducing inelastic regeneration by a factor of 10 (kaons scattered inelastically may be assigned to the wrong beam). Downstream of its regenerator each beam is a coherent superposition of K_{L}^0 and K_{S}^0 mesons. The 2π decay rate is given by (5.9).

The phases $\phi_{\varrho_c} - \phi_{+-}$ and $\Delta\phi \equiv \phi_{00} - \phi_{+-}$ were extracted from the measured decay rates into both neutral and charged pions [36] (see Fig. 5.11). From the fits, performed simultaneously to both regenerators, they found

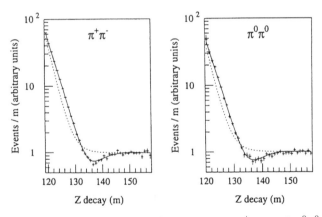

Fig. 5.11. Measured rates for decay into $\pi^+\pi^-$ and $\pi^0\pi^0$. The predictions from the fits with (*solid line*) and without (*dotted line*) the interference term are also shown [36]

---

[27] Since they used their own values for τ_{S} and $\Delta m_{\mathrm{L,S}}$, derived assuming (5.12), they could not report ϕ_{+-} in the same fit.

$$\phi_{+-} = (43.53 \pm 0.97)^\circ, \quad \phi_{00} - \phi_{+-} = (0.62 \pm 1.03)^\circ. \tag{5.16}$$

As with the E731 data, the interference terms were fitted by $\cos\left[t\Delta m_{\mathrm{L,S}} - (\phi_{+-} - \phi_{\varrho_{\mathrm{c}}})\right]$ for $\pi^+\pi^-$ and $\cos\left[t\Delta m_{\mathrm{L,S}} - (\phi_{+-} - \phi_{\varrho_{\mathrm{c}}} + \Delta\phi)\right]$ for $\pi^0\pi^0$ events.

5.3 Comparison of NA31 and E731 Experimental Techniques

We conclude this chapter with a few brief comments regarding the experiments NA31 and E731/E773 (see also [37]).

The presence of a regenerator in E731 leads to the quantum-mechanical interference between the $K_{\mathrm{L}}^0 \to 2\pi$ and $K_{\mathrm{S}}^0 \to 2\pi$ amplitudes, the measurement of which can provide independent confirmation of an ε'/ε signal.

The NA31 experiment had to be concerned with possible shifts in the overall detection efficiency, since the $K_{\mathrm{L}}^0 \to 2\pi$ and $K_{\mathrm{S}}^0 \to 2\pi$ decays were collected at different times under different rate conditions.

The K_{S}^0 decay distribution was not uniform in E731, resulting in large relative acceptance corrections.

Because of the shorter decay region, the residual background from $3\pi^0$ decays in the $K_{\mathrm{L}}^0 \to 2\pi^0$ sample in E731 is considerably smaller than in NA31.

The energy and position resolutions of the NA31 electromagnetic calorimeter are superior to those of E731. However, the plane resolution of its tracking chambers is much worse. As a consequence, the background in the $K_{\mathrm{L}}^0 \to \pi^+\pi^-$ sample is significantly smaller in E731 (NA31 was forced to discard about 40% of its $\pi^+\pi^-$ data in order to keep the $e\pi\nu$ background low).

Concerning backgrounds, it should be noted that a 1% shift in the double ratio (5.2) corresponds to 1.6×10^{-3} in ε'/ε. The largest backgrounds in E731 and NA31 were at a few-percent level.

In the Fermilab experiment the regenerator beam flux was significantly reduced by a 66 cm carbon absorber. Consequently, lack of statistics prevented them from extracting $\phi_{\varrho_{\mathrm{c}}}$ from the time-dependent charge asymmetry in semileptonic decays.

The presence of $\mathcal{D}(p)$ in (4.42) is a fundamental deficiency in this class of experiments, as a source of uncertainty in the NA31 data analysis.

6. Neutral Kaons
in Proton–Antiproton Annihilations

Proton–antiproton annihilations were first used as a source of neutral kaons in early 1960s (see, e.g., Armenteros et al. [38]). In a typical experiment of this kind, J. Steinberger and his collaborators produced K^0 and $\bar{K}^0$ mesons in equal, but relatively small, numbers in the annihilation of low-energy antiprotons in a liquid-hydrogen chamber at Brookhaven:

$$\bar{p} + p \to \bar{K}^0 K^+ + \text{pions}, \quad \bar{p} + p \to K^0 K^- + \text{pions}. \tag{6.1}$$

Due to strangeness conservation in strong interactions, the K^0 ($\bar{K}^0$) is "tagged" by the charge sign of the accompanying kaon.[28] Figure 6.1 shows the time distribution of the semileptonic decays $K^0(\bar{K}^0) \to \ell^\pm \pi^\mp \nu$ measured by Franzini et al. [38].

6.1 The CPLEAR Experiment at CERN

High-precision studies of CP violation based on this idea began a quarter of a century later at CERN, following the construction of the Low Energy Antiproton Ring (LEAR) and a dedicated detector. The CPLEAR experiment produces intense fluxes of tagged K^0 and $\bar{K}^0$ mesons by stopping low-energy antiprotons (200 MeV/c, 10^6 antiprotons/second) from LEAR in a low-density hydrogen target:

$$(\bar{p}p)_{\text{rest}} \to \bar{K}^0 K^+ \pi^-, \quad (\bar{p}p)_{\text{rest}} \to K^0 K^- \pi^+. \tag{6.2}$$

The branching ratio for each of the above two processes is about 0.2%, which means that K^0 and $\bar{K}^0$ mesons are produced in equal numbers. However, the tagging efficiencies for K^0 and $\bar{K}^0$ are not identical because of different cross-sections for interactions of K^+ and K^- mesons in the detector material. Note also that K^0 and $\bar{K}^0$ undergo coherent regeneration in the detector, which must be taken into account.

Tagged K^0 and $\bar{K}^0$ beams offer the possibility of observing directly K^0_L-K^0_S interference. The $K^0(\bar{K}^0)$ decay rate to any final state f reads (see (4.39), (4.40))

---

[28] Reactions (6.1) have also been used to test charge conjugation invariance in strong interactions.

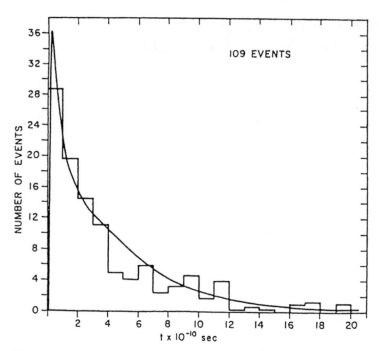

Fig. 6.1. The result from Franzini et al. [38] on the time-distribution of the leptonic decays of an equal mixture of K^0 and $\bar{K}^0$ mesons. The solid curve is the prediction of the $\Delta S = \Delta Q$ rule

$$\Gamma_{K^0, \bar{K}^0 \to f} = \frac{1}{2}(1 \mp 2\mathrm{Re}\,\epsilon)\Gamma_{K_S^0 \to f}\Big\{|\eta_f|^2 e^{-\Gamma_L t} + e^{-\Gamma_S t}$$

$$\pm 2|\eta_f| e^{-(\Gamma_S + \Gamma_L)t/2}\,\cos(\delta_m t - \phi_f)\Big\},\qquad(6.3)$$

where $\Gamma_{K_S^0 \to f} \equiv |\langle f \mid \hat{H}_w \mid K_S^0\rangle|^2$; as before, the upper (lower) signs refer to K^0 $(\bar{K}^0)$.

As for (4.6), the time-dependent decay asymmetry is defined by

$$A_f(t) \equiv \frac{\Gamma\left(\bar{K}^0 \to f\right) - \Gamma\left(K^0 \to f\right)}{\Gamma\left(\bar{K}^0 \to f\right) + \Gamma\left(K^0 \to f\right)}.\qquad(6.4)$$

Using (6.3), $A_f(t)$ can be expressed as

$$A_f(t) \approx 2\mathrm{Re}\,\epsilon - \frac{2|\eta_f| e^{(\Gamma_S - \Gamma_L)t/2}\,\cos(\delta_m t - \phi_f)}{1 + |\eta_f|^2 e^{(\Gamma_S - \Gamma_L)t}},\qquad(6.5)$$

thereby isolating the K_L^0-K_S^0 interference term in (6.3). We see that:

(a) since $K^0 \to f$ and $\bar{K}^0 \to f$ are CP-conjugate processes, any difference in their rates is a clear sign of CP violation;

(b) by measuring $\mathcal{A}_f(t)$, one can determine both the modulus and the phase of the CP-violating parameter η_f;

(c) the acceptances common to K^0 and $\bar{K}^0$ decays cancel in $\mathcal{A}_f(t)$.

The CPLEAR detector is shown in Fig. 6.2. It consists of: a spherical gaseous hydrogen target (16 bar pressure); cylindrical tracking chambers; a threshold Čerenkov counter (filled with liquid freon and sandwiched between two layers of scintillator) to identify charged kaons; and an electromagnetic calorimeter consisting of lead plates interspaced with streamer tubes. All components are assembled inside a solenoidal magnet, which provides a uniform 0.44 tesla field neeeded to measure the momenta of charged particles.

The acceptance-corrected decay rates of initially tagged K^0 and $\bar{K}^0$ mesons into $\pi^+\pi^-$ are shown in Fig. 6.3. The CP violating parameter η_{+-} was determined from the time-dependent asymmetry in the decay rates: expression (6.5) was fitted to the observed asymmetry in Fig. 6.4, keeping $|\eta_{+-}|$ and ϕ_{+-} as free parameters, with the result [39a] (in the fit they used their own value of $\Delta m_{\mathrm{L,S}}$ obtained from semileptonic decays of neutral kaons):

$$|\eta_{+-}| = (2.312 \pm 0.054) \times 10^{-3}, \quad \phi_{+-} = (42.7 \pm 1.8)^\circ. \tag{6.6}$$

See also [39b] for a correlation analysis of η_{+-} and $\Delta m_{\mathrm{L,S}}$ based on the results from different experiments, which yielded $\langle \phi_{+-} \rangle = 43.82^\circ \pm 0.63^\circ$.

Unlike in most other experiments, the measurement of $\Delta m_{\mathrm{L,S}}$ by CPLEAR is independent of ϕ. The K^0_{L}-K^0_{S} mass difference can be extracted from the decay rate asymmetry

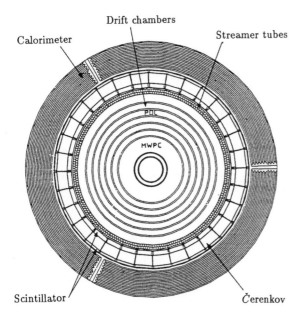

Fig. 6.2. The layout of the CPLEAR detector at CERN

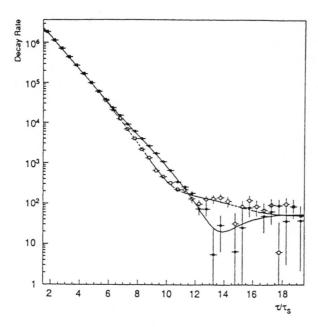

Fig. 6.3. Acceptance-corrected decay rate of K^0 ($\square$) and $\bar{K}^0$ ($\bullet$) into $\pi^+\pi^-$. The lines are the expected rates [39a]

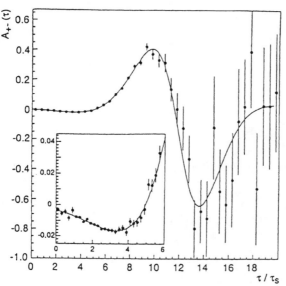

Fig. 6.4. Decay rate asymmetry versus the proper decay time. The solid line is the result of the fit described in [39a]. The inset displays the data at the short decay times with a refined binning

$$\mathcal{A}_{\Delta m}(t) \equiv \frac{[\Gamma_+(t) - \Gamma_-(t)] - [\overline{\Gamma}_+(t) - \overline{\Gamma}_-(t)]}{[\Gamma_+(t) + \Gamma_-(t)] + [\overline{\Gamma}_+(t) + \overline{\Gamma}_-(t)]}, \tag{6.7}$$

where the K^0 and $\bar{K}^0$ decay rates

$$\Gamma_+(t) \equiv \Gamma\left(K^0 \to \ell^+ \pi^- \nu\right), \quad \Gamma_-(t) \equiv \Gamma\left(K^0 \to \ell^- \pi^+ \bar{\nu}\right),$$
$$\overline{\Gamma}_+(t) \equiv \Gamma\left(\bar{K}^0 \to \ell^+ \pi^- \nu\right), \quad \overline{\Gamma}_-(t) \equiv \Gamma\left(\bar{K}^0 \to \ell^- \pi^+ \bar{\nu}\right) \tag{6.8}$$

and t is the decay eigentime. The expression for $\mathcal{A}_{\Delta m}(t)$ follows readily from (4.20)

$$\begin{aligned}
\mathcal{A}_{\Delta m}(t) &\equiv \frac{4\left(1 - |x|^2\right) e^{-\Gamma t} \cos(\delta_m t)}{2|1 + x|^2 e^{-\Gamma_S t} + 2|1 - x|^2 e^{-\Gamma_L t}} \\
&\approx \frac{2e^{-\Gamma t} \cos(\delta_m t)}{(1 + 2\mathrm{Re}\ x) e^{-\Gamma_S t} + (1 - 2\mathrm{Re}\ x) e^{-\Gamma_L t}},
\end{aligned} \tag{6.9}$$

where we neglected the terms proportional to $|x|^2$ and used $x + x^* = 2\mathrm{Re}\ x$. A possible violation of the $\Delta S = \Delta Q$ rule is taken into account by the parameter $\mathrm{Re}\ x$.

The measured asymmetry is shown in Fig. 6.5. Leaving $\Delta m_{\mathrm{L,S}}$ and $\mathrm{Re}\ x$ as free parameters, they fitted (6.9) to the data in Fig. 6.5 and obtained [40]

$$\Delta m_{\mathrm{L,S}} = (0.5274 \pm 0.0029) \times 10^{10}\,\mathrm{s}^{-1}, \tag{6.10}$$

thus confirming the low value of $\Delta m_{\mathrm{L,S}}$ measured by E731.

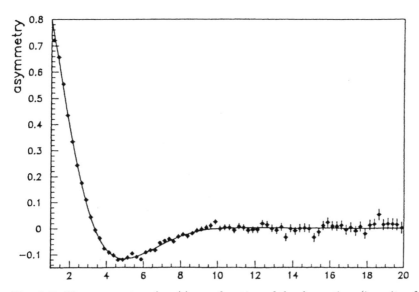

Fig. 6.5. The asymmetry $\mathcal{A}_{\Delta m}(t)$ as a function of the decay time (in units of τ_s). The solid line represents the result of the fit [40]

Their most recent results on $\Delta m_{\mathrm{L,S}}$ and Re x (full statistics) are [41a]

$$\Delta m_{\mathrm{L,S}} = (0.5295 \pm 0.00202) \times 10^{10}\,\mathrm{s}^{-1},$$

$$\mathrm{Re}\ x = [-1.8 \pm 4.1(\mathrm{stat}) \pm 4.5(\mathrm{syst})] \times 10^{-3}. \tag{6.11}$$

As explained in Appendix A, the K_{S}^0 meson may decay into the kinematics-supressed and CP-allowed final state $\pi^+\pi^-\pi^0$ with $L = l = 1$ and $CP = +1$, or into the kinematics-favored and CP-forbidden state $\pi^+\pi^-\pi^0$ with $L = l = 0$ and $CP = -1$. This results in a Dalitz plot distribution which is symmetric with respect to the π^+ and π^- for the CP-violating amplitude and antisymmetric for the CP-conserving amplitude. Thus by integrating the decay amplitude over the entire phase space of the $K^0 \to \pi^+\pi^-\pi^0$ decay, the CP-allowed contribution can be eliminated.

From (4.38), and defining

$$\eta_{+-0} \equiv \frac{A_{K_{\mathrm{S}}^0 \to \pi^+\pi^-\pi^0}}{A_{K_{\mathrm{L}}^0 \to \pi^+\pi^-\pi^0}}, \tag{6.12}$$

the decay amplitude for $K^0(\bar{K}^0) \to \pi^+\pi^-\pi^0$ can be expressed as

$$A_{+-0} = A_{K_{\mathrm{L}}^0 \to \pi^+\pi^-\pi^0} \frac{1}{\sqrt{2}} e^{-im_{\mathrm{L}}t}$$

$$\times \left\{ (1 \mp \epsilon)e^{-\Gamma_{\mathrm{L}}t/2} \pm (1 \mp \epsilon)\eta_{+-0}e^{(i\delta_{\mathrm{m}} - \Gamma_{\mathrm{S}}/2)t} \right\}, \tag{6.13}$$

where the upper (lower) signs refer to K^0 ($\bar{K}^0$). The corresponding decay rates read

$$\Gamma_{+-0} \propto (1 \mp 2\mathrm{Re}\ \epsilon) \left\{ e^{-\Gamma_{\mathrm{L}}t} + |\eta_{+-0}|^2 e^{-\Gamma_{\mathrm{S}}t} \right.$$

$$\left. \pm e^{-(\Gamma_{\mathrm{S}}+\Gamma_{\mathrm{L}})t/2} \left[\eta_{+-0}^* e^{-i\delta_{\mathrm{m}}t} + \eta_{+-0}e^{i\delta_{\mathrm{m}}t} \right] \right\}. \tag{6.14}$$

The time-dependent decay rate asymmetry, which is a direct measure of the K_{L}^0-K_{S}^0 interference, is given by

$$\mathcal{A}_{+-0}(t) \equiv \frac{\overline{\Gamma}_{+-0}(t) - \Gamma_{+-0}(t)}{\overline{\Gamma}_{+-0}(t) + \Gamma_{+-0}(t)}$$

$$\approx 2\mathrm{Re}\ \epsilon - 2e^{-(\Gamma_{\mathrm{S}}-\Gamma_{\mathrm{L}})t/2}$$

$$\times \left\{ \mathrm{Re}\ \eta_{+-0}\cos(\delta_{\mathrm{m}}t) - \mathrm{Im}\ \eta_{+-0}\sin(\delta_{\mathrm{m}}t) \right\}, \tag{6.15}$$

where $\overline{\Gamma}$ and Γ are the $\bar{K}^0$ and K^0 decay rates, respectively. A recent result on the CP-violating parameter η_{+-0}, based on the full statistics of CPLEAR, is [42a]

$$\mathrm{Re}\ \eta_{+-0} = \left[-2 \pm 7(\mathrm{stat})^{+4}_{-1}(\mathrm{syst}) \right] \times 10^{-3},$$

$$\mathrm{Im}\ \eta_{+-0} = \left[-2 \pm 9(\mathrm{stat})^{+2}_{-1}(\mathrm{syst}) \right] \times 10^{-3}. \tag{6.16}$$

Additional information about their measurement of η_{+-0} can be found elsewhere [42b].

The Fermilab experiment E621 has also published a result on Im η_{+-0} by fixing Re $\eta_{+-0} = \mathrm{Re}\ \epsilon$ and assuming *CPT* invariance [43]:

$$\mathrm{Im}\ \eta_{+-0} = [-15 \pm 30] \times 10^{-3}. \tag{6.17}$$

6.2 Is *CP* Violation Compensated by Time-Reversal Asymmetry?

If *CPT* is conserved, the observed *CP* violation demonstrates the failure of time-reversal invariance. A direct test of T asymmetry in the K^0 system was suggested by Aharony and Kabir in 1970 [44]. As shown in Appendix D, the operation of time reversal gives the identity

$$\langle K^0\ |\ \mathrm{e}^{-i\hat{H}t}\ |\ \bar{K}^0\rangle = \langle \bar{K}^0\ |\ \mathrm{e}^{-i\hat{H}_T t}\ |\ K^0\rangle \tag{6.18}$$

where $\hat{H}_T$ is the time-reverse of the hamiltonian $\hat{H}$. If $\hat{H}_T = \hat{H}$, the two amplitudes are equal. A nonzero value of the ratio of transition intensities

$$\mathcal{A}_T(t) \equiv \frac{\Gamma(\bar{K}^0 \to K^0) - \Gamma(K^0 \to \bar{K}^0)}{\Gamma(\bar{K}^0 \to K^0) + \Gamma(K^0 \to \bar{K}^0)}, \tag{6.19}$$

where

$$\begin{aligned}
\Gamma(\bar{K}^0 \to K^0) &\equiv \left|\langle K^0\ |\ \mathrm{e}^{-i\hat{H}t}\ |\ \bar{K}^0\rangle\right|^2, \\
\Gamma(K^0 \to \bar{K}^0) &\equiv \left|\langle \bar{K}^0\ |\ \mathrm{e}^{-i\hat{H}t}\ |\ K^0\rangle\right|^2,
\end{aligned} \tag{6.20}$$

would thus imply $\hat{H}_T \neq \hat{H}$, i.e., a violation of time-reversal invariance.

The first evidence for time-reversal noninvariance has been reported by the CPLEAR collaboration based on semileptonic decays of tagged neutral kaons [41b].They extracted

$$\mathcal{A}_T(t) = \frac{\overline{\Gamma}_+(t) - \Gamma_-(t)}{\overline{\Gamma}_+(t) + \Gamma_-(t)} \approx 4\mathrm{Re}\ \epsilon + 2\mathrm{Im}\ x\frac{\sin(\delta_m t)}{\cosh(\Delta\Gamma t) - \cos(\delta_m t)} \tag{6.21}$$

from the measured decay rates $\overline{\Gamma}_+(t) \equiv \Gamma\left[\bar{K}^0(t=0) \to \ell^+\pi^-\nu\right]$ and $\Gamma_-(t) \equiv \Gamma\left[K^0(t=0) \to \ell^-\pi^+\bar{\nu}\right]$ (see Fig. 6.6). Expression (6.21), which was derived assuming *CPT* invariance, follows readily from (4.13), with $\Delta\Gamma \equiv (\Gamma_S - \Gamma_L)/2$ (see also Appendix D).

The CPLEAR measurement yielded

$$\mathcal{A}_T(t) = [6.6 \pm 1.3(\mathrm{stat}) \pm 1.0(\mathrm{syst})] \times 10^{-3}, \tag{6.22}$$

which should be compared with $4\mathrm{Re}\ \epsilon \approx 6.6 \times 10^{-3}$ (see (6.21), (4.26), and (4.28)), assuming Im $x = 0$.

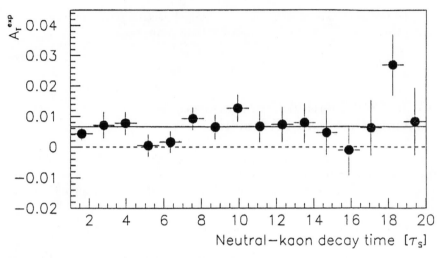

Fig. 6.6. The asymmetry $\mathcal{A}_T$, indicating a violation of T invariance in semileptonic decays of neutral kaons. The solid line represents the fitted average $\langle \mathcal{A}_T(t) \rangle$ [41b]

7. Neutral Kaons in Electron–Positron Annihilations

7.1 The DAΦNE Project

The DAΦNE project at Frascati (Italy) will study the process

$$e^+e^- \to \text{virtual photon} \to \Phi \to K^0\bar{K}^0, \quad J_\Phi^{PC} = 1^{--}, \tag{7.1}$$

by producing at rest about 5000 Φ mesons per second at the centre-of-mass energy $\sqrt{s} = 1020$ MeV. As explained in Sect. 1.4, the neutral kaon pair in (7.1) is in a pure $C = -1$ quantum state:

$$\begin{aligned}
|\Phi\rangle &= \frac{1}{\sqrt{2}} \left\{ |K^0(z)\bar{K}^0(-z)\rangle - |\bar{K}^0(z)K^0(-z)\rangle \right\} \\
&= \frac{1}{\sqrt{2}} \left\{ |K_{\mathrm{L}}^0(z)K_{\mathrm{S}}^0(-z)\rangle - |K_{\mathrm{S}}^0(z)K_{\mathrm{L}}^0(-z)\rangle \right\},
\end{aligned} \tag{7.2}$$

i.e., the final state is either $K^0\bar{K}^0 - \bar{K}^0K^0$ or $K_{\mathrm{L}}^0 K_{\mathrm{S}}^0 - K_{\mathrm{S}}^0 K_{\mathrm{L}}^0$. With an expected collider luminosity of 5×10^{32} cm^{-2} s^{-1}, about 10^{10} coherently produced $K_{\mathrm{L}}^0 K_{\mathrm{S}}^0$ pairs per year will provide a particularly beautiful method of quantum interferometry.

To study the time evolution of the state (7.2), we denote by $f_1(t_1, +z)$ and $f_2(t_2, -z)$ any two final states in the neutral-kaon decay, and define the amplitude ratios

$$\eta_i \equiv \frac{\langle f_i | \hat{H}_{\mathrm{w}} | K_{\mathrm{L}}^0 \rangle}{\langle f_i | \hat{H}_{\mathrm{w}} | K_{\mathrm{S}}^0 \rangle}. \tag{7.3}$$

Using (1.18)

$$|K_{\mathrm{S,L}}^0(t)\rangle = e^{-(im_{\mathrm{S,L}} + \Gamma_{\mathrm{S,L}}/2)t} |K_{\mathrm{S,L}}^0(t=0)\rangle \equiv e^{-i\mathcal{M}_{\mathrm{S,L}}t} |K_{\mathrm{S,L}}^0(t=0)\rangle \tag{7.4}$$

and (7.2), the decay amplitude for $\Phi \to f_1 f_2$ reads (see also [46a])

$$\begin{aligned}
A_{\Phi \to f_1 f_2} &= \frac{1}{\sqrt{2}} \Big\{ A_{\mathrm{L} \to f_1} A_{\mathrm{S} \to f_2} \, e^{-i(\mathcal{M}_{\mathrm{S}} t_2 + \mathcal{M}_{\mathrm{L}} t_1)} \\
&\quad - A_{\mathrm{S} \to f_1} A_{\mathrm{L} \to f_2} \, e^{-i(\mathcal{M}_{\mathrm{S}} t_1 + \mathcal{M}_{\mathrm{L}} t_2)} \Big\} \\
&= \frac{A_{\mathrm{S} \to f_1} A_{\mathrm{S} \to f_2}}{\sqrt{2}} e^{-i\mathcal{M}t/2} \left\{ \eta_1 \, e^{i\Delta\mathcal{M}\Delta t/2} - \eta_2 \, e^{-i\Delta\mathcal{M}\Delta t/2} \right\},
\end{aligned} \tag{7.5}$$

where

$$\mathsf{A}_{\mathrm{S,L}\to f} \equiv \langle f \mid \hat{H}_{\mathrm{w}} \mid K^0_{\mathrm{S,L}} \rangle, \quad \mathsf{A}_{\Phi\to f_1 f_2} \equiv \langle f_1 f_2 \mid \hat{H}_{\mathrm{w}} \mid \Phi \rangle, \tag{7.6}$$

and

$$\Delta t = t_2 - t_1, \quad t = t_1 + t_2, \quad \mathcal{M} = \mathcal{M}_{\mathrm{L}} + \mathcal{M}_{\mathrm{S}}, \quad \Delta\mathcal{M} = \mathcal{M}_{\mathrm{L}} - \mathcal{M}_{\mathrm{S}}. \tag{7.7}$$

For equal times ($t_1 = t_2$) the amplitude $\mathsf{A}_{\Phi\to f_1 f_2}$ vanishes for identical final states of the two kaons ($f_1 = f_2 = \pi^+\pi^-, \pi^0\pi^0$). This result follows from Bose statistics: since the kaon is spinless, the total angular momentum in a 2π state must be zero. Two identical 2π states thus behave like two identical spinless bosons, and therefore cannot be in an antisymmetric spatial state (recall that $P_\Phi = -1$).

However, if the two final states are not identical at equal times, e.g., $f_1 = \pi^+\pi^-$ and $f_2 = \pi^0\pi^0$, then

$$\mathsf{A}_{\Phi\to\pi^+\pi^-\pi^0\pi^0} \propto \eta_{+-} - \eta_{00} = 3\varepsilon' \quad \text{for} \quad \Delta t = 0. \tag{7.8}$$

Simultaneous decay to distinct CP eigenstates, such as $\pi^+\pi^-$ and $\pi^0\pi^0$, is therefore an unambiguous signature of direct CP violation.

The decay rate corresponding to (7.5) is given by

$$\Gamma_{\Phi\to f_1 f_2} = \frac{\Gamma_{\mathrm{S}\to f_1}\Gamma_{\mathrm{S}\to f_2}}{2}\, \mathrm{e}^{-\Gamma t}$$
$$\times \left\{ |\eta_1|^2 \mathrm{e}^{-\Delta\Gamma\Delta t} + |\eta_2|^2 \mathrm{e}^{\Delta\Gamma\Delta t} - 2|\eta_1||\eta_2|\cos\Theta \right\} \tag{7.9}$$

with

$$\Theta \equiv \delta_{\mathrm{m}}\Delta t + \phi_1 - \phi_2, \quad \Gamma \equiv (\Gamma_{\mathrm{S}} + \Gamma_{\mathrm{L}})/2, \quad \Delta\Gamma \equiv (\Gamma_{\mathrm{S}} - \Gamma_{\mathrm{L}})/2. \tag{7.10}$$

From an experimental point of view, it is convenient to consider the probability distribution in the relative time Δt:

$$\Gamma(f_1, f_2; \Delta t) \equiv \frac{1}{2} \int_{\Delta t}^{\infty} \mathrm{d}(t_1 + t_2)\, \Gamma_{\Phi\to f_1 f_2}. \tag{7.11}$$

In the above expression we integrate over all experimentally "accessible" times $t = t_1 + t_2$, keeping Δt constant. The factor $1/2$ is the Jacobian for the transformation from (t_1, t_2) to $(t, \Delta t)$. For $\Delta t \geq 0$ a straightforward calculation yields

$$\Gamma(\Delta t > 0) = \frac{\Gamma_{\mathrm{S}\to f_1}\Gamma_{\mathrm{S}\to f_2}}{4\Gamma} \tag{7.12}$$
$$\times \left\{ |\eta_1|^2 \mathrm{e}^{-\Gamma_{\mathrm{S}}\Delta t} + |\eta_2|^2 \mathrm{e}^{-\Gamma_{\mathrm{L}}\Delta t} - 2|\eta_1||\eta_2|\mathrm{e}^{-\Gamma\Delta t}\cos\Theta \right\}.$$

For $\Delta t < 0$ we obtain an expression analogous to this, with $\Delta t \to |\Delta t|$ and with indices 1 and 2 interchanged.

If we set $f_1 = \pi^+\pi^-$ and $f_2 = \pi^0\pi^0$ in (7.5), the quadratic terms in $|\mathsf{A}_{\Phi\to f_1 f_2}|^2$ are proportional to

$$|\eta_{+-}|^2 \approx |\varepsilon|^2 [1 + 2\mathrm{Re}\,(\varepsilon'/\varepsilon)], \quad |\eta_{00}|^2 \approx |\varepsilon|^2 [1 - 4\mathrm{Re}\,(\varepsilon'/\varepsilon)] \tag{7.13}$$

(see (5.4) and (5.5)) and the mixed terms to

Interference $\propto -\eta_{+-}(\eta_{00})^* e^{i\delta_m \Delta t} - (\eta_{+-})^* \eta_{00} e^{-i\delta_m \Delta t}$

$$\approx -2|\varepsilon|^2 \big\{ [1 - \mathrm{Re}\,(\varepsilon'/\varepsilon)]$$

$$\times \cos \delta_m \Delta t - 3\mathrm{Im}\,(\varepsilon'/\varepsilon) \sin \delta_m \Delta t \big\} . \qquad (7.14)$$

Expression (7.12) now reads

$$\Gamma(\Delta t > 0) = \mathcal{C}\,|\varepsilon|^2 \big\{ (1 + 2\mathfrak{R})e^{-\Gamma_S \Delta t} + (1 - 4\mathfrak{R})e^{-\Gamma_L \Delta t}$$

$$- 2e^{-\Gamma \Delta t}\,[(1 - \mathfrak{R})\cos\theta - 3\mathfrak{I}\sin\theta] \big\} , \qquad (7.15)$$

where

$$\mathcal{C} \equiv \frac{\Gamma_{S,00}\Gamma_{S,+-}}{4\Gamma}, \quad \mathfrak{R} \equiv \mathrm{Re}\,(\varepsilon'/\varepsilon), \quad \mathfrak{I} \equiv \mathrm{Im}\,(\varepsilon'/\varepsilon), \quad \theta \equiv \delta_m \Delta t. \quad (7.16)$$

For $\Delta t < 0$ there is an expression analogous to this with $\Delta t \to |\Delta t|$, $\Gamma_S \leftrightarrow \Gamma_L$ and $\mathrm{Im}\,(\varepsilon'/\varepsilon) \to -\mathrm{Im}\,(\varepsilon'/\varepsilon)$. A nonzero value of ε'/ε would thus result in an asymmetry between the $\Delta t > 0$ and $\Delta t < 0$ distributions. The interference patterns for different combinations of final states shown in Fig. 7.1 can be used to extract $\mathrm{Re}\,(\varepsilon'/\varepsilon)$, $\mathrm{Im}\,(\varepsilon'/\varepsilon)$, $\Delta m_{L,S}$, $|\eta_{\pi\pi}|$, $\phi_{\pi\pi}$, etc., as discussed in [46b, c]. The asymmetry in $K_L^0 \to \ell^\pm \pi^\mp \nu$ decays provides tests of T and CPT invariance.

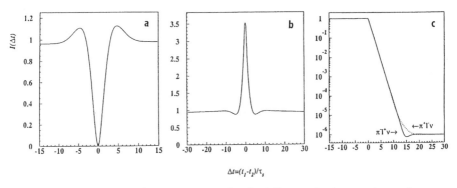

Fig. 7.1. Calculated interference patterns for the following final states in two-kaon decays: **(a)** $f_1 = \pi^+\pi^-$, $f_2 = \pi^0\pi^0$; **(b)** $f_1 = l^+\pi^-\nu$, $f_2 = l^-\pi^+\bar{\nu}$; **(c)** $f_1 = 2\pi$, $f_2 = l\pi\nu$

Measuring ε'/ε at DAΦNE requires very accurate reconstruction of t_{+-} and t_{00}. The high-statistics data from $K_L^0 \to \ell^\pm \pi^\mp \nu$, 3π decays will be used to map detector acceptance and reconstruction efficiency. About 10^7 $\pi^+\pi^-\pi^0\pi^0$ decays are needed for a statistical error of 10^{-4} on $\mathrm{Re}\,(\varepsilon'/\varepsilon)$, the CP-violating $K_L^0 \to 2\pi$ decays being the modes with the limiting statistics. In this context, the K_S^0 and K_L^0 decay lengths from $\Phi \to K_S^0 K_L^0$ are $\Lambda_S = \gamma\beta c\tau_S = 0.592$ cm and $\Lambda_L = \gamma\beta c\tau_L = 343$ cm, respectively. About one quarter of K_L^0 mesons are expected to decay within the tracking volume of the KLOE detector at DAΦNE.

The DAΦNE project is, undoubtedly, very versatile. It provides a novel method of quantum interferometry and precision tests of the discrete symmetries C, P and T not readily achieved in other experiments. The neutral kaons in the reaction (7.1) change their identity continually and in a completely correlated way. This can be used to test quantum mechanics by studying correlations of the Einstein-Podolsky-Rosen type (see Sect. 1.4 and [47, 48]). Note also that a Φ factory is characterized by low background, since about a third of the Φ-decay final states are neutral-kaon pairs $(K^+K^- : K^0\bar{K}^0 : \varrho\pi = 0.49 : 0.34 : 0.13)$.

8. Neutral Kaons in Fixed-Target Experiments

8.1 The Experiments KTeV and NA48

Presently there are two major fixed-target experiments with neutral kaons: NA48 at CERN and KTeV at Fermilab. The elegance and sophistication of these experiments reflect years of experience with K^0 beams, especially that gained with their predecessors, NA31 and E731, respectively. The main aim of each of the two groups is to determine $\mathrm{Re}\,(\varepsilon'/\varepsilon)$ with a precision of $\approx 10^{-4}$. In order to achieve this goal it is necessary to collect roghly 4 million $K_L^0 \to 2\pi^0$ decays, an increase of about an order of magnitude compared with NA31 or E731.

The KTeV apparatus is shown in Fig. 8.1. The experiment retains the basic features of E731: it uses two beams, which are side by side and identical in shape, and records all four decay modes simultaneously. To reduce inelastic regeneration, an "active" regenerator made of plastic scintillator is employed. The most significant improvement with respect to E731 is the use of an

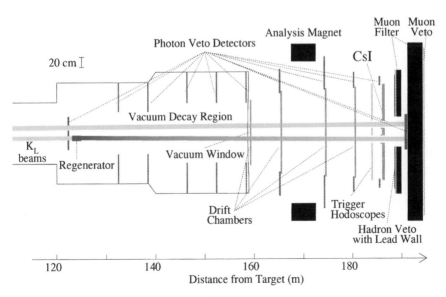

Fig. 8.1. A schematic drawing of the KTeV detector

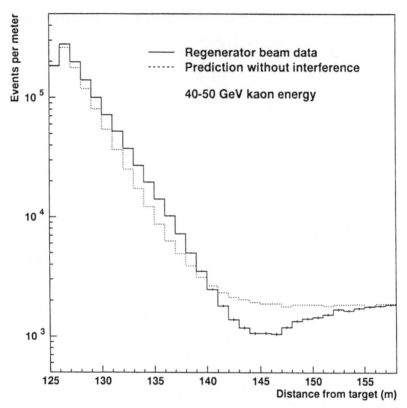

Fig. 8.2. Interference in $K^0 \to \pi^+\pi^-$ decays measured by KTeV

undoped CsI electromagnetic calorimeter. This should result in a much better energy resolution and thus more efficient background rejection in the $2\pi^0$ decay channel. KTeV expects to achieve an error of $0.5°$ in the determination of ϕ_{+-} and ϕ_{00}, and also of ϕ_ϱ using semileptonic decays. Fig. 8.2 shows their preliminary result on kaon interference.

The NA48 detector is sketched in Fig. 8.3. An important element of the new design is the concurrent presence of (almost) collinear K^0_L and K^0_S beams in the detector, produced by protons hitting two different targets. A channeling crystal is used to simultaneously attenuate and deflect protons emerging from the K^0_L target, which are then sent toward the K^0_S target located close to the detector (see Fig. 8.4 [49]). The protons producing the K^0_S component are tagged in order to distinguish between the K^0_S and K^0_L beams. A major improvement compared with NA31 is a liquid-krypton calorimeter with superior energy and position resolution. A magnetic spectrometer has been added for charged decay modes.

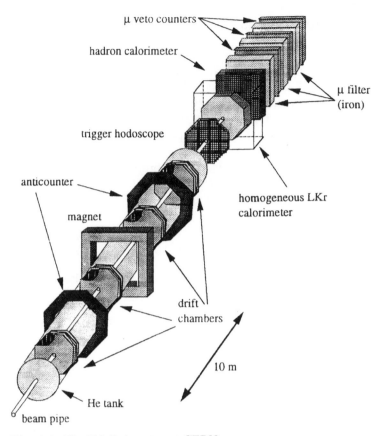

Fig. 8.3. The NA48 detector at CERN

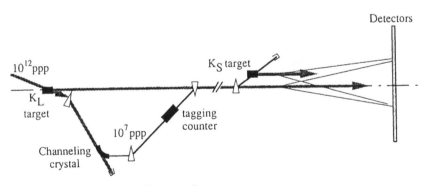

Fig. 8.4. Simultaneous K_L^0 and K_S^0 beams in NA48

Supplement: Kaon Beams

As a supplement to the preceding section, we will describe (a) the production of K mesons by a high-energy proton beam striking a stationary target, and (b) the techniques of electrostatic and RF separation of K^+ beams. Our discussion is partly based on the lecture notes by B. Winstein [104]. Other sources of K-mesons are discussed in Sects. 6.1 (proton–antiproton annihilations) and 7.1 (e^+e^- collisions). An experiment that uses K^+ "beams" at rest is described in Sect. 9.3.1.

The production of charged kaons by incident protons on a Be target can be expressed as a yield per incident proton per GeV/c kaon momentum per steradian of the solid angle Ω. An approximate parametrization of the yield at angle θ is given by

$$\frac{\mathrm{d}^2\sigma}{\mathrm{d}p\,\mathrm{d}\Omega}\left[\frac{\mathrm{mb}}{(\mathrm{GeV/c})\mathrm{str}}\right] \propto \frac{p(1-x)^a}{(1+p_t^2)}\left(1+5\,\mathrm{e}^{-20x}\right) \equiv p\,f(x,p_t) \tag{S8.1}$$

with $x = p/p_{\mathrm{beam}}$ and $p_t = \theta p$. Using $\mathrm{d}^3p = p^2\mathrm{d}p\,\mathrm{d}\Omega$, we can write

$$\frac{\mathrm{d}^2\sigma}{\mathrm{d}p\,\mathrm{d}\Omega} = p^2\frac{\mathrm{d}^3\sigma}{\mathrm{d}^3p} \approx p\left(E\,\frac{\mathrm{d}^3\sigma}{\mathrm{d}^3p}\right) \equiv p\,\sigma_{\mathrm{inv}}, \tag{S8.2}$$

where "inv" denotes Lorentz invariance. The kaon yield per incident proton therefore reads

$$\mathrm{Yield} \equiv \frac{\mathrm{d}^2N}{\mathrm{d}p\,\mathrm{d}\Omega} = \frac{p\,\sigma_{\mathrm{inv}}}{\sigma_{\mathrm{inel}}}, \tag{S8.3}$$

where σ_{inel} is the inelastic cross-section. From (S8.1) we see that the kaon production is most copious in the forward direction ($p_t = 0$). The most significant difference in the production of K^+ and K^- mesons is in the x dependence:

$$K^+ \propto (1-x)^3, \quad K^- \propto (1-x)^6. \tag{S8.4}$$

By simple quark-counting, one can show that the relative production rates of charged and neutral kaons are given by

$$K^0 \sim (K^+ + K^-)/2, \quad \bar{K}^0 \sim K^-, \tag{S8.5}$$

i.e.,

$$\frac{K^0 - \bar{K}^0}{K^0 + \bar{K}^0} \approx \frac{K^+ - K^-}{K^+ + 3K^-}. \tag{S8.6}$$

The asymmetry in the K^0 and $\bar{K}^0$ production spectra is due to an interplay between associated production and baryon number conservation. Expressions (S8.5) and (S8.6) are corroborated by a measurement of the "dilution factor"

$$\mathcal{D}(p) \equiv \frac{K^0(p) - \bar{K}^0(p)}{K^0(p) + \bar{K}^0(p)} \tag{S8.7}$$

at the CERN SPS (the NA31 collaboration [32]; see Sect. 5.1).

Therefore, a proton beam produces a mixture of K^0 and $\bar{K}^0$ mesons that depends upon energy and angle. At high energies (large x), the K^0 production dominates. As was mentioned in Sect. 4.1, if K^0 and $\bar{K}^0$ mesons were produced with equal spectra ($\mathcal{D}(p) = 0$), the interference term in (4.42) could not be observed in a target experiment! Sufficiently far from the production target, where only the K_L^0 component can survive, the initial composition of the kaon beam becomes irrelevant.

In the forward direction ($\theta = 0°$), the neutral beam is dominated by photons and neutrons. Photons, which originate mainly from π^0 decays, are easily removed with a lead converter near the production target. It is sufficient to have a converter that is about 10 radiation lengths thick ($\approx 6\,\mathrm{cm}$), since the probability of photon non-conversion is $\mathrm{e}^{-(7/9)\times 10} \approx 4 \times 10^{-4}$. The charged products of the resulting electromagnetic shower are swept away by a magnetic field.

The neutrons are produced with the invariant cross-section

$$\sigma_{\mathrm{inv}}(n) \approx \frac{25\,\mathrm{mb}}{(\mathrm{GeV/c})^2}\, \mathrm{e}^{-5p_t}, \tag{S8.8}$$

which does not depend on $x = p/p_{\mathrm{beam}}$. The neutron flux is many times the kaon flux at $\theta = 0°$, and thus presents a much more serious problem than the photon beam. Fortunately, the kaon to neutron ratio can be enhanced very efficiently by targeting away from $\theta = 0°$. It follows from the above differential distributions that an enhancement factor of over 100 can be obtained with less than a factor of 2 loss in the kaon flux at $\theta \approx 5\,mr$ (see [104]).

In the design of a K^0 beam, one also has to consider: (a) the "soft" component of neutral particles in and around the beam ("beam halo"), (b) the forward hadronic "jet" and (c) the beam of noninteracting particles. The first component can be reduced considerably by collimators that subtend little solid angle compared with Ω_{beam}. To dispose of the other two components, a "beam dump" is used that is thick enough to completely absorb the hadronic shower, and is also sufficiently close to the target to reduce the muon flux originating from π and K decays.

As an example, we outline the main features of the KTeV beamline. The kaons are produced by 800 GeV protons striking a BeO target. The proton beam is delivered every minute in spills lasting 23 s. About half of the 3×10^{12} protons per spill interact in the target, while the other half are absorbed in a beam dump. Neutral kaons produced in the horizontal plane at an angle of 4.8 mrad relative to the proton direction point toward the detector. Two identical sets of collimators produce two secondary neutral beams at $\pm 0.8\,\mathrm{mrad}$ in the horizontal plane. The size of the beams is about $8\,\mathrm{cm} \times 8\,\mathrm{cm}$ at $180\,\mathrm{m}$ from the target. A large sweeping magnet downstream of the beam dump removes muons from both the primary target and the dump. The photon flux is reduced using a 7.6-cm-long lead absorber that transmits 55% of kaons and neutrons.

To further enhance the kaon to neutron ratio, a 52-cm-long Be absorber is used. To see how this enhancement works, note that the neutron and kaon total cross-sections have the following atomic number dependences:

$$\sigma_{\text{tot}}(n) \approx 49\,mb \times A^{0.77}, \quad \sigma_{\text{tot}}(K) \approx 24\,mb \times A^{0.84}, \quad A \geq 7. \quad (\text{S8.9})$$

The cross-section ratio is thus greatest for small A. The number of interaction lengths, X_{int}, in the Be absorber ($A = 9$, $L = 52\,\text{cm}$) is

$$X_{\text{int}} = \sigma_{\text{tot}} \left(\frac{N_0 \varrho}{A} \right) L = \begin{cases} 1.74, & \text{neutrons,} \\ 0.97, & \text{kaons,} \end{cases} \quad (\text{S8.10})$$

where $N_0 \approx 6 \times 10^{23}$ is Avogadro's number and $\varrho = 1.85\,\text{g/cm}^3$ is the density of Be. The respective transmission coefficients are $e^{-X_{\text{int}}} = 0.18$ (neutrons) and 0.38 (kaons), resulting in an enhancement factor of about 2.

As mentioned in Sect. 9.3, the purity of K^+ beams at Brookhaven (experiment E787) has been considerably improved through electrostatic (DC) separation. This method of particle separation works in the following way. An almost parallel beam enters a separator which has a vertical electric field and a horizontal magnetic field. The separator is tuned in such a way that the action of the magnetic field cancels the action of the electric field for the kaons, whereas the pions are deflected. At a vertical focus downstream of the seperator, the pion beam is displaced from the axis and then stopped. The separation is given by

$$\text{Separation (in rad)} \approx \frac{EL}{\text{p}} \left(\frac{1}{\beta_\pi} - \frac{1}{\beta_k} \right), \quad (\text{S8.11})$$

where, in the case of E787, $E = 60\,\text{KV/cm}$ and $L = 220\,\text{cm}$ (p is the beam momentum).

The method of RF separation employs, in general, two cavities and a system of quadrupole magnets between them. The first RF cavity imposes a transverse momentum kick on the beam of a few MeV/c per meter of cavity length. The second cavity, located downstream of the first one, is so tuned that the π^+ mesons arrive with the same phase that they had in the first cavity. Since the π^+ momenta have been reversed in the quadrupoles, the positive pions end up with no net kick. If the RF frequency and the distance between the cavities are such that the π^+s and protons are 360° out of phase at the second cavity, the protons will also receive no net kick. The K^+ mesons, on the other hand, are 90° out of phase with respect to the π^+s, and thus get a net transverse kick corresponding to $\approx 1\,\text{mrad}$. A beam plug downstram of the second cavity stops the pions and protons.

RF-separated K^+ beams can be used, for example, to search for the rare decay $K^+ \rightarrow \pi^+ \nu \bar{\nu}$ (see Sect. 9.3). Alternatively, they can be focused on a target to produce K^0 mesons by charge exchange.

9. The K^0 System in the Standard Model

9.1 Calculation of Δm_k and ϵ_k

So far our description of the K^0 system has been restricted to its quantum-mechanical properties and phenomenological implications of the observed CP violation. In this section we consider K^0-$\bar{K}^0$ mixing within the framework of gauge theories of the electroweak and strong interactions (the Standard Model (SM)). We will outline the calculation of the off-diagonal elements of the K^0-$\bar{K}^0$ mass matrix that generate the tiny K_L^0-K_S^0 mass difference and the CP-violating parameter ϵ (see (3.30) and (3.82)). This is the forerunner of all the calculations that are forbidden in the lowest order of the electroweak theory.

The smallness of the K_L^0-K_S^0 mass difference ($\Delta m_k = 3.5 \times 10^{-6}\,\mathrm{eV}$) enables the quantum-mechanical interference effects between the K_S^0 and K_L^0 components of a K^0 beam to spread out over macroscopic distances (see Sect. 1.5). Phenomenologically, it played an important role in the establishment of the *charm hypothesis*, as explained in what follows.

There is overwhelming experimental evidence that strangeness-changing neutral currents are heavily suppressed. For example,

$$\frac{\Gamma(K_L^0 \to \mu^+\mu^-)}{\Gamma(K_L^0 \to \text{all})} \approx 10^{-8}, \quad \frac{\Gamma(K^\pm \to \pi^\pm \nu\bar{\nu})}{\Gamma(K^\pm \to \text{all})} \approx 10^{-10}. \tag{9.1}$$

The suppression of $K_L^0 \to \mu^+\mu^-$ was very surprising because the analogous charged decay $K^+ \to \mu^+\nu_\mu$ has a branching ratio of 63% (see Fig. 9.1, where the mesons are represented, in accordance with SM, as bound states of quark–antiquark pairs). To explain this puzzle, Glashow, Iliopoulos and Maiani (GIM) incorporated the *charm* quark into the hadronic weak current originally proposed by Cabibbo. The weak charged current in the GIM model reads [50]

$$\mathcal{J}_\alpha^{CC} = \frac{g}{\sqrt{2}}\,(\bar{u}\,\bar{c})_L\,\gamma_\alpha \begin{pmatrix} d' \\ s' \end{pmatrix}_L, \tag{9.2}$$

where

$$\begin{pmatrix} d' \\ s' \end{pmatrix}_L = \begin{pmatrix} \cos\theta_c & \sin\theta_c \\ -\sin\theta_c & \cos\theta_c \end{pmatrix} \begin{pmatrix} d \\ s \end{pmatrix}_L, \quad \psi'(d',s') = V\,\psi(d,s), \tag{9.3}$$

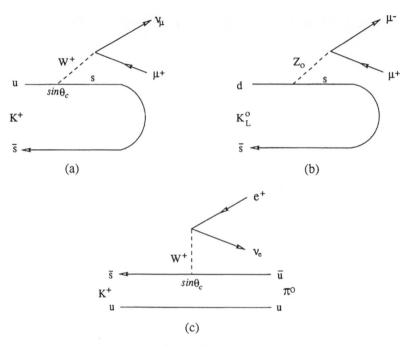

Fig. 9.1. Allowed decays $K^+ \to \mu^+\nu_\mu$ **(a)** and $K^+ \to \pi^0 e^+\nu_e$ **(c)**; forbidden first-order contribution to $K_L^0 \to \mu^+\mu^-$ **(b)**

g is the SU(2) weak coupling, γ_α are Dirac matrices and θ_c is the Cabibbo angle. The subscript L denotes "left-handed" spinors: $q_L \equiv \frac{1}{2}(1 - \gamma_5)q$. The quarks have charges $Q_{u,c,t} = +2/3$ and $Q_{d,s,b} = -1/3$. Since the *flavor* quantum numbers are not conserved in weak interactions, the *weak eigenstates d'* and *s'* are mixtures of the *mass eigenstates d* and s.[29] The mixing matrix V, which also couples the *up* (u) as well as the *charm* (c) quark to a linear combination of the *down* (d) and *strange* (s) quarks, is *unitary*:

$$V^\dagger V = V^{-1}V = \mathbb{1} \tag{9.4}$$

that is to say, the inverse of V is its hermitian conjugate ($\mathbb{1}$ is the unit matrix). The unitarity of V ensures the absence of strangeness-changing neutral currents. Indeed,

$$\overline{\psi'}\hat{O}\psi' = \overline{\psi}V^\dagger\hat{O}V\psi = \overline{\psi}\hat{O}V^\dagger V\psi = \overline{\psi}\hat{O}\psi \tag{9.5}$$

for any arbitrary operator $\hat{O}$. From (9.5) we conclude that the neutral component of the hadronic weak current does not contain terms that mix quark flavours. Changes of flavor always involve change of charge (hence $\Delta S = \Delta Q$). This explains why the amplitude in Fig. 9.1b is zero.

---

[29] It is sufficient to mix ("rotate") either (ds) or (uc).

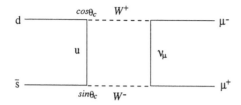

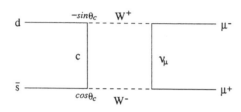

Fig. 9.2. Second-order contributions to $K_L^0 \to \mu^+ \mu^-$

The decay $K_L^0 \to \mu^+ \mu^-$ can also proceed through the diagrams in Fig. 9.2. The amplitudes corresponding to these diagrams are proportional to $\cos\theta_c \sin\theta_c$ and $-\sin\theta_c \cos\theta_c$, respectively. The second amplitude cancels most of the first. If the u and c quarks had the same mass, the cancellation would be exact (see (9.13)).

The foregoing discussion demonstrates the importance of the Cabibbo angle θ_c, which not only suppresses the weak transitions that are proportional to $\theta_c \approx 13°$, but also removes strangeness-changing neutral currents via the charm quark. The origin of θ_c, however, is not explained in the Standard Model.

It is instructive to calculate the amplitude that corresponds to the lowest-order diagrams contributing to Δm_k (see Fig. 9.3a,b). The lowest order for which the transition $s\bar{d} \leftrightarrow \bar{s}d$ is possible is fourth order in the weak hamiltonian $\hat{H}_w$: the two W bosons in Fig. 9.3a,b are emitted and reabsorbed, so that a total of four weak "vertices" are involved. Of the two sets of diagrams in Fig. 9.3a,b it is sufficient to compute the first one: it can be readily verified that, in the limit of vanishing external momenta, the amplitude corresponding to the second set is identical. We ignore, for the moment, the contribution of the *top* (t) quark, which is justified when computing Δm_k (but not ϵ), as explained later on. Neglecting the external quark momenta,[30] the electroweak Feynman rules applied to the *box diagrams* in Fig. 9.3a yield, in the so-called t'Hooft-Feynman gauge,[31]

---

[30] In the K^0 rest frame, their components are of the order of m_k and thus can be neglected compared with the W boson and heavy quark masses.

[31] The presence of unphysical scalars in this gauge can be ignored when the top quark is not included, since their coupling to fermions (f) is proportional to $m_f/m_w \ll 1$.

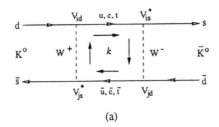

(a)

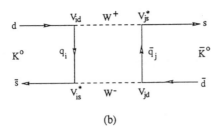

(b)

Fig. 9.3a,b. Feynman diagrams for the $\Delta S = 2$ transition $\bar{s}d \to sd$

$$\begin{aligned}
\mathrm{i}\mathcal{M} = {} & \left(\frac{-\mathrm{i}g}{\sqrt{2}}\sin\theta_c\right)^2 \left(\frac{-\mathrm{i}g}{\sqrt{2}}\cos\theta_c\right)^2 \\
& \times \int \frac{\mathrm{d}^4k}{(2\pi)^4}\left[\bar{u}_s\gamma_\lambda L\left(\frac{\mathrm{i}(\hat{k}+m_u)}{k^2-m_u^2}-\frac{\mathrm{i}(\hat{k}+m_c)}{k^2-m_c^2}\right)\gamma_\varrho L u_d\right. \\
& \times \bar{v}_s\gamma_\alpha L\left(\frac{\mathrm{i}(\hat{k}+m_u)}{k^2-m_u^2}-\frac{\mathrm{i}(\hat{k}+m_c)}{k^2-m_c^2}\right) \\
& \left. \times \gamma_\sigma L v_d\left(\frac{-\mathrm{i}g^{\lambda\sigma}}{k^2-m_w^2}\right)\left(\frac{-\mathrm{i}g^{\alpha\varrho}}{k^2-m_w^2}\right)\right]
\end{aligned} \tag{9.6}$$

($\hat{k} \equiv \gamma^\mu k_\mu$). The terms in u and c have opposite signs because of the GIM mechanism (see Fig. 9.2). We can simplify the above expression by using $\gamma_\mu L = R\gamma_\mu$, $L^2 = L$, $R^2 = R$ and $RL = LR = 0$ ($L, R \equiv (1 \mp \gamma_5)/2$). The quark masses in the numerators thus drop out, with the result

$$\mathrm{i}\mathcal{M} = g^4 \frac{\sin^2\theta_c\cos^2\theta_c}{4}(m_c^2-m_u^2)^2 \int \frac{\mathrm{d}^4k}{(2\pi)^4}\frac{k_\mu k_\nu}{a^2b^2c^2}T^{\mu\nu}, \tag{9.7}$$

where $a \equiv k^2 - m_u^2$, $b \equiv k^2 - m_c^2$, $c \equiv k^2 - m_w^2$ and

$$T^{\mu\nu} \equiv \bar{u}_s\gamma_\lambda\gamma^\mu\gamma_\varrho L u_d\bar{v}_s\gamma^\varrho\gamma^\nu\gamma^\lambda L v_d. \tag{9.8}$$

To calculate the integral

$$I_{\mu\nu} = \int \frac{\mathrm{d}^4k}{(2\pi)^4}\frac{k_\mu k_\nu}{(k^2-m_u^2)^2(k^2-m_c^2)^2(k^2-m_w^2)^2} \tag{9.9}$$

we use Feynman's "parameter formula"

$$\frac{1}{a^2 b^2 c^2} = 120 \int_0^1 x dx \int_0^{1-x} dy \, \frac{(1-x-y)y}{[a + (b-a)x + (c-a)y]^6} \tag{9.10}$$

and

$$\int d^4 k \, \frac{k_\mu k_\nu}{(k^2 + t)^n} = i\pi^2 \frac{\Gamma(n-3)}{2\Gamma(n)} \frac{g_{\mu\nu}}{t^{n-3}} \quad n \geq 4. \tag{9.11}$$

Both expressions are frequently employed in the evaluation of loop integrals.
Writing $a + (b-a)x + (c-a)y \approx k^2 - [m_w^2 y + (m_c^2 - m_u^2)x + m_u^2] \equiv k^2 + t$, we have

$$\begin{aligned}
I_{\mu\nu} &= 120 \int_0^1 x \, dx \int_0^{1-x} dy \, (1-x-y)y \int \frac{d^4 k}{(2\pi)^4} \frac{k_\mu k_\nu}{(k^2 + t)^6} \\
&= \frac{-i g_{\mu\nu}}{16\pi^2 m_w^6} \int_0^1 x \, dx \int_0^{1-x} dy \, \frac{(1-x-y)y}{[y + x(m_c^2 - m_u^2)m_w^{-2}]^3} \\
&= \frac{-i g_{\mu\nu}}{64\pi^2(m_c^2 - m_u^2)m_w^4} \tag{9.12}
\end{aligned}$$

since $m_w^2 \gg m_c^2 \gg m_u^2$ [$m_w = (80.3 \pm 0.05)\,\mathrm{GeV}/c^2$, $m_c = (1.3 \pm 0.3)\,\mathrm{GeV}/c^2$]. Therefore,

$$\begin{aligned}
i\mathcal{M} &= \frac{-i g^4}{2^8 \pi^2 m_w^2} \frac{m_c^2 - m_u^2}{m_w^2} \sin^2\theta_c \cos^2\theta_c \\
&\quad \times [\bar{u}_s \gamma_\lambda \gamma_\nu \gamma_\varrho L u_d \bar{v}_s \gamma^\varrho \gamma^\nu \gamma^\lambda L v_d]. \tag{9.13}
\end{aligned}$$

The factor $(m_c^2 - m_u^2)/m_w^2$ represents the typical GIM suppression mentioned above (it contains m_w^{-2} because the loop integration is cut off by m_w). A single box diagram would yield $\mathcal{M} \propto g^4 m_w^{-2}$.

The Dirac algebra in (9.13) can be simplified by virtue of

$$\begin{aligned}
\gamma_\lambda \gamma_\nu \gamma_\varrho &= (g_{\lambda\nu} g_{\varrho\sigma} + g_{\lambda\sigma} g_{\nu\varrho} - g_{\lambda\varrho} g_{\nu\sigma} + i\varepsilon_{\lambda\nu\varrho\sigma} \gamma_5) \gamma^\sigma, \\
\gamma^\varrho \gamma^\nu \gamma^\lambda &= (g^{\varrho\nu} g^{\lambda\alpha} + g^{\varrho\alpha} g^{\nu\lambda} - g^{\varrho\lambda} g^{\nu\alpha} + i\varepsilon^{\varrho\nu\lambda\alpha} \gamma_5) \gamma_\alpha \tag{9.14}
\end{aligned}$$

($\varepsilon_{0123} = +1, \varepsilon^{0123} = -1$). It is then straightforward to show that

$$\gamma_\lambda \gamma_\nu \gamma_\varrho (1 - \gamma_5) \otimes \gamma^\varrho \gamma^\nu \gamma^\lambda (1 - \gamma_5) = 4 \gamma^\alpha (1 - \gamma_5) \otimes \gamma_\alpha (1 - \gamma_5), \tag{9.15}$$

where the symbol $\otimes$ separates the matrices from two different fermion lines. Expressing the SU(2) weak coupling g in terms of the Fermi coupling constant G_F ($g^2/8m_w^2 = G_F/\sqrt{2}$), it follows ($m_c^2 \gg m_u^2$) that

$$\mathcal{M} = \frac{-G_F^2}{8\pi^2} m_c^2 \sin^2\theta_c \cos^2\theta_c [\bar{u}_s \gamma^\alpha (1 - \gamma_5) u_d \bar{v}_s \gamma_\alpha (1 - \gamma_5) v_d]. \tag{9.16}$$

The Feynman amplitude (9.16) describes the $\Delta S = 2$ transitions $s\bar{d} \leftrightarrow \bar{s}d$, with quark–antiquark pairs in the intermediate states. What we are actually concerned with are the hadronic transitions $K^0 \leftrightarrow \bar{K}^0$ mediated by genuine physical states. We thus express the above amplitude (obtained in the limit of vanishing external quark momenta) as the matrix element of an equivalent four-fermion operator between the K^0 and $\bar{K}^0$ states (normalized to unity in a volume V):

$$M_{12} = \frac{\langle \bar{K}^0 | \hat{H}_{\mathrm{w}}^{\mathrm{eff}} | K^0 \rangle}{2m_k} \tag{9.17}$$

$$\equiv \frac{-G_{\mathrm{F}}^2}{16\pi^2} \frac{m_c^2}{2m_k} \sin^2\theta_{\mathrm{c}} \cos^2\theta_{\mathrm{c}} \langle \bar{K}^0 | \bar{s}\gamma^\alpha(1-\gamma_5)d\, \bar{s}\gamma_\alpha(1-\gamma_5)d | K^0 \rangle.$$

In deriving this expression the spinors in (9.16) were replaced by the field operators $\hat{\psi}_{s,d} \equiv s, d$. Also, a factor of $1/2$ was included to compensate for the fact that $\hat{H}_{\mathrm{w}}^{\mathrm{eff}}$ contains four terms which contribute to $s\bar{d} \leftrightarrow \bar{s}d$, two corresponding to Fig. 9.3a and two to Fig. 9.3b, as can be readily verified by writing the fields in terms of creation and annihilation operators.

This brings us to the most obscure part in the calculation of M_{12}, the evaluation of the hadronic matrix element. Following the early calculations of Δm_k based on the box diagrams in Fig. 9.3a,b [51a], it is customary to insert the vacuum intermediate state in the middle of the four-fermion operator, which amounts to neglecting strong-interaction effects. Since the renormalized operator $\hat{H}_{\mathrm{w}}^{\mathrm{eff}}$ cannot really be treated as a product of two factors, the whole procedure is dubious, to say the least. These uncertainties are embodied in the parameter

$$\mathcal{B}_K \equiv \frac{\langle \bar{K}^0 | \bar{s}\gamma^\alpha(1-\gamma_5)d | 0 \rangle \langle 0 | \bar{s}\gamma_\alpha(1-\gamma_5)d | K^0 \rangle}{\langle \bar{K}^0 | \bar{s}\gamma^\alpha(1-\gamma_5)d\, \bar{s}\gamma_\alpha(1-\gamma_5)d | K^0 \rangle}. \tag{9.18}$$

Using the definition of the $K_{2\ell}$ decay constant f_k [32],

$$\langle 0 | \bar{s}\gamma_\alpha\gamma_5 d | K^0(q) \rangle \equiv f_k q_\alpha, \quad f_k^{\mathrm{exp}} \approx 160\,\mathrm{MeV}, \tag{9.19}$$

we obtain, in the K^0 rest frame,

$$\langle \bar{K}^0 | \bar{s}\gamma^\alpha(1-\gamma_5)d | 0 \rangle \langle 0 | \bar{s}\gamma_\alpha(1-\gamma_5)d | K^0 \rangle = \frac{8}{3}(f_k m_k)^2. \tag{9.20}$$

The presence of the additional factor of $8/3$ is explained in Appendix F (see also [98]). Therefore,

$$\Delta m_k = -2\,\mathrm{Re}\,M_{12} = \frac{G_{\mathrm{F}}^2}{6\pi^2} f_k^2\, \mathcal{B}_K m_k m_c^2 \sin^2\theta_{\mathrm{c}} \cos^2\theta_{\mathrm{c}}, \tag{9.21}$$

which is in good agreement with the measured value of Δm_k, provided $\mathcal{B}_K \approx 1$ and $m_c \approx 1.5\,\mathrm{GeV}/c^2$. Historically, a correct upper limit was set on the charm mass in this way [51b] before *charmonium* (the bound state of a charm–anticharm pair) was observed. In the light of what we said earlier, however, this must be viewed as a fortuitous coincidence.

---

[32] In the matrix element of a vector-axial vector current between the vacuum and a pseudoscalar state only the axial current contributes (see (9.105)).

Since the K^0-$\bar{K}^0$ mass splitting is caused by weak interactions, one would expect Δm_k to be comparable to the K_S^0 decay width, which is indeed the case (see (1.41)).

The quarks inside the K mesons are "glued" together by the strong force, resulting in *gluonic* corrections to the box diagrams of Fig. 9.3a,b. It is beyond the scope of this book to discuss strong-interaction effects on Δm_k (see [54] for an extensive review of the subject), except to mention that among various nonperturbative calculations of the denominator in (9.18), the lattice gauge theory yields the most accurate value: $\mathcal{B}_K = 0.8 \pm 0.2$. We also note that the real part of M_{12} is dominated by low momenta ($k < m_c$) in the loop diagram. In this region the effect of "virtual" low-energy transitions $K^0 \rightarrow \pi\pi, \pi, \eta, \varrho, \eta' \rightarrow \bar{K}^0$ is important, yet difficult to estimate reliably.

We now turn to the evaluation of the *CP*-violating parameter ϵ based on the box diagrams in Fig. 9.3a,b. According to (3.82),

$$\epsilon = \frac{-\,e^{i\phi_{\pi\pi}}}{\sqrt{2}\,\Delta m_k}\,\mathrm{Im}\,M_{12}. \tag{9.22}$$

With its two *weak isospin doublets* (ud') and (cs'), the GIM model cannot account for an imaginary part of the matrix element M_{12} (see below). This motivated Kobayashi and Maskawa (KM) to introduce, in 1973, a third quark doublet. Their proposal is not a trivial extension of the four-quark model because it allows for the existence of *CP* violation within SM. The KM model [52] was proposed before the discovery of even the charm quark (see Chap. 1). In this context, recall that a third quark *generation* is not required to explain K^0-$\bar{K}^0$ mixing.

The GIM model can be extended to include the two additional quarks b (*bottom*, or *beauty*) and t (*top*) by defining, in analogy with (9.2) and (9.3),

$$\mathcal{J}_\alpha^{CC} = \frac{g}{\sqrt{2}}\,(\,\bar{u}\,\bar{c}\,\bar{t}\,)_L\,\gamma_\alpha\,\mathbf{V}\begin{pmatrix} d \\ s \\ b \end{pmatrix}_L, \quad \mathbf{V}^\dagger = \mathbf{V}^{-1}, \tag{9.23}$$

where $\mathbf{V}$ is the unitary, 3×3 Cabibbo–Kobayashi–Maskawa (CKM) matrix. The matrix elements of $\mathbf{V}$ can be expressed in terms of a certain number of independent parameters. A unitary $n \times n$ matrix for n quark generations is characterized by $n_\theta = (n-1)n/2$ rotation angles and $n_\delta = (n-1)(n-2)/2$ physical phases. For $n = 2$ we have $n_\delta = 0$ and $n_\theta = 1$, the only parameter being the Cabibbo angle θ_c. For three generations $n_\theta = 3$ and $n_\delta = 1$, i.e., in addition to three mixing angles there is also a nonvanishing phase δ. Therefore, the GIM matrix is real and has only one parameter, whereas the CKM matrix is complex and contains four independent parameters.

There is a number of ways (three dozen, in fact) to express the elements of $\mathbf{V}$ in terms of three rotation angles and one phase. The parametrization suggested by Wolfenstein [53] is particularly convenient because it emphasizes the observed strong hierarchy of the CKM matrix elements:

$$\mathbf{V} \equiv \begin{pmatrix} V_{ud} & V_{us} & V_{ub} \\ V_{cd} & V_{cs} & V_{cb} \\ V_{td} & V_{ts} & V_{tb} \end{pmatrix}$$

$$\approx \begin{pmatrix} 1 - \lambda^2/2 & \lambda & A\lambda^3 r e^{-i\delta} \\ -\lambda(1 + A^2\lambda^4 r e^{i\delta}) & 1 - \lambda^2/2 - A^2\lambda^6 r e^{i\delta} & A\lambda^2 \\ A\lambda^3(1 - r e^{i\delta}) & -A\lambda^2(1 + \lambda^2 r e^{i\delta}) & 1 \end{pmatrix}, \qquad (9.24)$$

where A and r are of the order of unity and $\lambda \equiv \sin\theta_c = 0.22 \pm 0.002$. The notation for V_{ij} on the left in (9.24) may seem peculiar: the matrix element V_{ud}, for example, does not refer to the mixing of u and d quarks (they cannot mix because of charge conservation), but rather to that of the d' and d states (see (9.3)). Recall, however, that $\mathbf{V}$ may also be viewed as containing the transition amplitudes for weak processes, in which case V_{ij} represents the relative strength of the transition $i \leftrightarrow j$. The rows and columns of $\mathbf{V}$ must satisfy $\sum_j |V_{ij}|^2 = \sum_i |V_{ij}|^2 = 1$.

From (9.24) we infer that (a) the quarks of one generation are coupled to those of the successive generations with decreasing strength: $V_{ub} \ll V_{us} < V_{ud}$, etc.; (b) $\mathbf{V}$ is almost diagonal, i.e., it is practically the unit matrix; (c) $|V_{ij}| \approx |V_{ji}|$; and (d) the third generation contributes marginally to the 2×2 GIM submatrix (otherwise the Cabibbo model would not have been so successful).

Before outlining the calculation of ϵ, we would like to stress that the Standard Model does not reveal the origin of *CP* violation — it merely allows for its existence. Indeed, by offering no insight into the CKM matrix, the model betrays one of its most conspicuous shortcomings. The hope is that *CP* violation will contribute to a more profound understanding of *flavor dynamics*, this intricate component of SM that involves spontaneous breaking of electroweak symmetry and contains most of the free parameters of the model.

Neglecting the external quark momenta, the amplitude corresponding to the box diagram in Fig. 9.3a reads[33]

$$i\mathcal{M} = \frac{g^4}{4} \int \frac{d^4k}{(2\pi)^4} \sum_{ij} \xi_i \xi_j$$

$$\times \left(\bar{u}_s \gamma_\lambda L \frac{i(\hat{k} + m_i)}{k^2 - m_i^2} \gamma_\varrho L u_d \right) \left(\bar{v}_s \gamma_\alpha L \frac{i(\hat{k} + m_j)}{k^2 - m_j^2} \gamma_\sigma L v_d \right)$$

$$\times \frac{-i(g^{\lambda\sigma} - k^\lambda k^\sigma/m_w^2)}{k^2 - m_w^2} \frac{-i(g^{\alpha\varrho} - k^\alpha k^\varrho/m_w^2)}{k^2 - m_w^2}, \qquad (9.25)$$

---

[33] This time we employ the "unitary gauge", in which there are no diagrams with unphysical scalars. Since $m_t > m_w$, the terms $k^\lambda k^\sigma/m_w^2$ and $k^\alpha k^\varrho/m_w^2$ in (9.25) cannot be neglected.

where

$$\xi_i \equiv V_{is}^* V_{id}. \tag{9.26}$$

As explained earlier, the quark masses in the numerators drop out. Furthermore, the unitarity of the CKM matrix (which implies that any pair of rows or columns are orthogonal)

$$(V^\dagger V)_{ij} = \delta_{ij} = (VV^\dagger)_{ij} \leftrightarrow \sum_k V_{ki}^* V_{kj} = \delta_{ij} = \sum_k V_{ik} V_{jk}^* \tag{9.27}$$

gives

$$\sum_{k=u,c,t} \xi_k = \xi_u + \xi_c + \xi_t \equiv \sum_{k=u,c,t} V_{ks}^* V_{kd} = 0 \tag{9.28}$$

($\sum_k \xi_k = 0$ is an off-diagonal element of the unit matrix), thus simplifying the calculation considerably. Writing

$$\frac{1}{(k^2 - m_i^2)(k^2 - m_j^2)} = \frac{m_i^2 m_j^2}{k^4 (k^2 - m_i^2)(k^2 - m_j^2)} - \frac{1}{k^4}$$
$$+ \frac{1}{k^2(k^2 - m_i^2)} + \frac{1}{k^2(k^2 - m_j^2)} \tag{9.29}$$

we see that (a) the last three terms do not contribute because of (9.28), and (b) if the quark masses were equal, the GIM cancellation would result in $\mathcal{M} = 0$. The remaining term yields a convergent integral that is straightforward to evaluate. For $m_u = 0$ it follows that

$$i\mathcal{M} = \frac{g^4}{16} \int \frac{\mathrm{d}^4 k}{(2\pi)^4} \frac{1}{k^2(k^2 - m_w^2)^2}$$
$$\times \left[\frac{\xi_c^2 m_c^4}{(k^2 - m_c^2)^2} + \frac{\xi_t^2 m_t^4}{(k^2 - m_t^2)^2} + \frac{2\xi_c \xi_t m_c^2 m_t^2}{(k^2 - m_c^2)(k^2 - m_t^2)} \right]$$
$$\times \left(1 - \frac{2k^2}{m_w^2} + \frac{k^4}{4m_w^4} \right) \bar{u}_d \gamma_\beta (1 - \gamma_5) u_s \bar{v}_d \gamma^\beta (1 - \gamma_5) v_s, \tag{9.30}$$

where we used $k^\mu k^\nu = g^{\mu\nu} k^2/4$ (which holds because of symmetric integration in $\mathrm{d}^4 k$), $\hat{k}\hat{k} = k^2$ and expression (9.15). Replacing the spinors by the corresponding field operators, equation (9.30) can be rewritten as (see the text between (9.16) and (9.21))

$$M_{12} = \frac{-iG_F^2 f_k^2 m_k}{12\pi^4}$$
$$\times \int \frac{\mathrm{d}^4 k}{k^2} \left[\frac{\xi_c m_c^4}{(k^2 - m_t^2)^2} + \frac{\xi_t^2 m_t^4}{(k^2 - m_t^2)^2} + \frac{2\xi_c \xi_t m_c^2 m_t^2}{(k^2 - m_c^2)(k^2 - m_t^2)} \right]$$
$$\times \left[1 - \frac{3k^4}{4(k^2 - m_w^2)^2} \right]. \tag{9.31}$$

We will ignore, for the moment, the term $3k^4/4(k^2 - m_w^2)^2$ in the second square bracket (it gives corrections that are of the order of m_q^2/m_w^2 and therefore negligible except for the top quark: $m_t = (180 \pm 15)\,\text{GeV}/c^2$. The remaining integrals are easy to solve, resulting in

$$M_{12} \approx \frac{-G_F^2 f_k^2 m_k}{12\pi^2} \mathcal{B}_K \left[\xi_c^2 m_c^2 + \xi_t^2 m_t^2 + 2\xi_c \xi_t m_c^2 \ln(m_t^2/m_c^2) \right]. \qquad (9.32)$$

The first, second and third terms correspond to box diagrams with $c\bar{c}$, $t\bar{t}$ and $c\bar{t} + t\bar{c}$ loops, respectively.

The CKM matrix elements given in (9.24) yield ($re^{i\delta} \equiv \varrho + i\eta$)

$$\xi_c^2 \approx \lambda^2 \left(1 - 2iA^2\lambda^4\eta \right),$$
$$\xi_t^2 \approx A^4\lambda^{10} \left[(1 - \varrho)^2 - \eta^2 + 2i\eta(1 - \varrho) \right], \qquad (9.33)$$
$$2\xi_c\xi_t \approx 2A^2\lambda^6 (1 - \varrho + i\eta).$$

Clearly, the real part of M_{12} is dominated by $\xi_c^2 \approx \sin^2\theta_c$, despite the fact that $m_t \gg m_c$. This justifies our earlier claim that the top quark can be ignored as far as Δm_k is concerned.

Regarding the imaginary part of M_{12}, note that the contributions from the three terms in (9.32), which are of the same order, are suppressed by the common factor $A^2\lambda^6 \sin\delta$. This "explains" why the CP-violating parameter ϵ is so small, and shows explicitly that ϵ can be attributed to the presence of a complex phase factor in the CKM matrix.

From (9.22), (9.32) and (9.33) it follows that

$$\epsilon \approx e^{i\phi_{\pi\pi}} \frac{G_F^2 f_k^2 m_k}{6\sqrt{2}\,\pi^2 \Delta m_k} \mathcal{B}_K A^2\lambda^6\eta$$
$$\times \left\{ m_c^2 \left[\ln(m_t^2/m_c^2) - 1 \right] + m_t^2 A^2\lambda^4(1 - \varrho) \right\}. \qquad (9.34)$$

The result of a more detailed calculation of ϵ, which includes strong-interaction (QCD) corrections to the lowest-order electroweak amplitude, can be expressed as [54] (the original calculation is due to Inami and Lim [55])

$$|\epsilon| = \frac{G_F^2 f_k^2 m_k m_w^2}{12\sqrt{2}\,\pi^2 \Delta m_k} \mathcal{B}_K$$
$$\times \left\{ \eta_c x_c \text{Im}\,\xi_c^2 + 2\eta_{ct} E(x_c, x_t) \text{Im}\,\xi_c\xi_t + \eta_t E(x_t) \text{Im}\,\xi_t^2 \right\}$$
$$= \mathcal{C}_\epsilon \mathcal{B}_K A^2\lambda^6\eta \left\{ [\eta_{ct} E(x_c, x_t) - \eta_c x_c] + A^2\lambda^4 \eta_t E(x_t)(1 - \varrho) \right\}. \qquad (9.35)$$

Here $x_i = m_i^2/m_w^2$, the coefficients η_i are perturbative QCD corrections ($\eta_c = 1.38$, $\eta_t = 0.57$, $\eta_{ct} = 0.47$) and

$$\mathcal{C}_\epsilon \equiv \frac{G_F^2 f_k^2 m_k m_w^2}{6\sqrt{2}\,\pi^2 \Delta m_k} = 3.8 \times 10^4 \quad \left(G_F = 1.17 \times 10^{-5}\,\text{GeV}^{-2} \right). \qquad (9.36)$$

The functions $E(x_t)$ and $E(x_c, x_t)$, obtained after the loop integration in (9.31), depend weakly on the top quark mass:[34]

---

[34] When next-to-leading-order QCD corrections are included, the "current–quark" mass $\tilde{m}_t(m_t) = m_t^*[1 - (4/3)\alpha_s(m_t)/\pi]$ should be used (m_t^* is the pole mass measured in collider experiments).

$$E(x_t) = x_t \left[\frac{4 - 11x_t + x_t^2}{4(1 - x_t)^2} - \frac{3x_t^2 \ln x_t}{2(1 - x_t)^3} \right]$$

$$E(x_c, x_t) = x_c \left[\ln(x_t/x_c) - \frac{3x_t}{4(1 - x_t)} - \frac{3x_t^2 \ln x_t}{4(1 - x_t)^2} \right]. \tag{9.37}$$

Since Im M_{12} is dominated by large loop momenta ($k > m_c$), the low-energy mesonic transitions $K^0 \to \pi\pi, \pi, \eta, \varrho, \eta' \to \bar{K}^0$, while significant for Δm_k, do not affect ϵ. As explained in [54], η_c depends strongly on the QCD parameter Λ, a number that is not fixed by the theory (one can think of Λ as the energy at which quasi-free quarks and gluons bind themselves together to become hadrons). Luckily, $\eta_c x_c$ is smaller than either of the other two terms in (9.35).

9.2 B^0-$\bar{B}^0$ Mixing and Constraints on CKM Parameters

The computed value of ϵ depends on the CKM parameters A, η and ϱ (λ is known to a high precision). From (9.24),

$$A = \frac{|V_{cb}|}{\lambda^2}, \quad \sqrt{\varrho^2 + \eta^2} = \frac{1}{\lambda} \left| \frac{V_{ub}}{V_{cb}} \right|. \tag{9.38}$$

Combining measurements of inclusive semileptonic B decays to charmed mesons with those of the exclusive decays $B^0 \to (D^{*+})D^+\ell^-\bar{\nu}$, it is possible to deduce [56]

$$|V_{cb}| = 0.038 \pm 0.002 \to A = 0.79 \pm 0.04. \tag{9.39}$$

The ratio $|V_{ub}/V_{cb}|$ can be obtained from semileptonic decays of B mesons, produced on the $\Upsilon(4S)$ resonance, by measuring the lepton energy spectrum above the endpoint allowed for the predominant $B \to D\ell\nu$ transitions. There are large theoretical uncertainties in the predicted lepton spectrum used to extract V_{ub}. We quote [56]

$$\left| \frac{V_{ub}}{V_{cb}} \right| = 0.08 \pm 0.015 \to \mathfrak{r}_b \equiv \sqrt{\varrho^2 + \eta^2} = 0.36 \pm 0.07. \tag{9.40}$$

The magnitude of V_{td} places an additional constraint on ϱ and η. $|V_{td}|$ can be determined from "virtual" processes involving $t \leftrightarrow d$ transitions. The only available process is B^0-$\bar{B}^0$ mixing, identified by the presence of "same-sign" leptons in semileptonic decays of $B^0 \bar{B}^0$ pairs:

$$e^+e^- \to \Upsilon(4S) \to B^0\bar{B}^0 \to \ell^\pm\ell^\pm + \text{anything}. \tag{9.41}$$

This reaction is possible if one of the neutral B-mesons can transform to its antiparticle before decaying. Otherwise, the flavor-specific transitions $B^0 \to \ell^-\nu X$ and $\bar{B}^0 \to \ell^+\bar{\nu}X$ would result in opposite-sign dilepton events in (9.41).

It does not require much imagination, nor effort, to extend our description of K^0-$\bar{K}^0$ mixing to the B^0 system. The Feynman diagrams responsible for

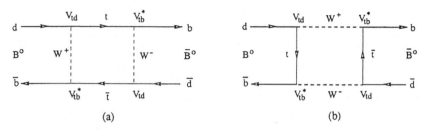

Fig. 9.4a,b. Box diagrams for B^0-$\bar{B}^0$ mixing

B^0-$\bar{B}^0$ oscillation are shown in Fig. 9.4a,b. Whereas the top quark contribution to kaon mixing is suppressed by $|V_{td}V_{ts}^*|^2 \approx \lambda^{10}$, in the case of neutral B mesons[35]

$$\Delta m(B_d^0)_t \propto m_t^2 |V_{td}V_{tb}^*|^2 \approx m_t^2 A^2 \lambda^6 \left[(1-\varrho)^2 + \eta^2\right],$$
$$\Delta m(B_s^0)_t \propto m_t^2 |V_{ts}V_{tb}^*|^2 \approx m_t^2 A^2 \lambda^4. \tag{9.42}$$

Since $\Delta m(B_d^0)_c \propto m_c^2 \lambda^6$ and $\Delta m(B_s^0)_c \propto m_c^2 \lambda^4$, the diagrams with t quarks in both internal fermion lines dominate the B^0-$\bar{B}^0$ mass difference. The expressions for $\Delta m(B_{d,s}^0)$ follow directly from (9.31) and (9.37):

$$\Delta m(B_{q=d,s}^0) = \frac{G_F^2}{6\pi^2} [m_B f_B^2 \mathcal{B}_B]_{d,s} \eta_B$$
$$\times \int d^4 k \frac{|V_{tq}V_{tb}^*|^2 m_t^4}{k^2(k^2 - m_t^2)^2} \left[1 - \frac{3k^4}{4(k^2 - m_w^2)^2}\right]$$
$$= \frac{G_F^2}{6\pi^2} m_w^2 [m_B f_B^2 \mathcal{B}_B]_{d,s} |V_{tq}V_{tb}^*|^2 \eta_B E(x_t), \tag{9.43}$$

where $m_{B_d} = 5.28\,\mathrm{GeV}/c^2$, $m_{B_s} = 5.38\,\mathrm{GeV}/c^2$ and the QCD correction factor $\eta_B = 0.55 \pm 0.01$. The parameter $\mathcal{B}_B$, analogous to $\mathcal{B}_K$, is related to the probability of $(d\bar{b})$ or $(s\bar{b})$ quarks forming a B^0 meson. The product $f_B^2 \mathcal{B}_B$ contains all the uncertainties associated with the hadronic matrix element. From lattice QCD calculations, $f_{B_d}\sqrt{\mathcal{B}_{B_d}} = (200\pm40)\,\mathrm{MeV}$ (see [54]). Solving (9.43) for $|V_{td}V_{tb}^*|$ yields

$$|V_{td}V_{tb}^*| \approx |V_{td}| = (8.3 \pm 1.85) \times 10^{-3}, \tag{9.44}$$

where we used the world-average value [57]

$$\Delta m(B_d^0) = (3.0 \pm 0.13) \times 10^{-4}\,\mathrm{eV} \rightarrow \frac{\Delta m(B_d^0)}{\Delta m_k} = 85.8 \pm 3.6. \tag{9.45}$$

Since

$$\frac{1}{\lambda}\left|\frac{V_{td}}{V_{cb}}\right| = \sqrt{(\varrho-1)^2 + \eta^2}, \tag{9.46}$$

---

[35] In contrast to the K^0 system, there are two neutral B mesons: the $B_d^0(d\bar{b})$ and the $B_s^0(s\bar{b})$.

it follows that

$$\mathfrak{r}_t \equiv \sqrt{(\varrho - 1)^2 + \eta^2} = 0.99 \pm 0.22. \tag{9.47}$$

The constraints on ϱ and η given by (9.40) and (9.47) have the form of rings centred at $(0,0)$ and $(1,0)$ in the (ϱ,η) plane, with radii $\mathfrak{r}_b$ and $\mathfrak{r}_t$, respectively. Another constraint on these parameters comes from CP-violating decays of K_L^0 mesons, as given by (9.35). The allowed values of ϱ and η are shown in Fig. 9.5.

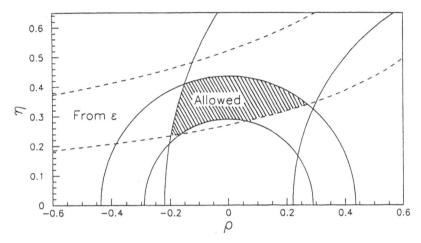

Fig. 9.5. Allowed region in the ρ-η plane

As mentioned earlier, the CKM matrix can be parametrized in many different ways. We will now demonstrate, however, that there exists a quantity that "measures" the amount of CP violation in a parametrization-independent manner, and in doing so will provide a simple geometrical interpretation of the unitarity equation $(VV^\dagger)_{ij} = \delta_{ij}$.

Expressions (9.27) lead to six relations that can be represented as triangles in the complex (ϱ,η) plane. Consider, for example, the unitarity relation

$$V_{ud}V_{td}^* + V_{us}V_{ts}^* + V_{ub}V_{tb}^* = 0. \tag{9.48}$$

Since $V_{ud} \approx V_{tb} \approx 1$, $V_{ts}^* \approx -V_{cb}$ and $V_{us} = \lambda$, we can write (9.48) as

$$\frac{V_{ub}}{V_{cb}} + \frac{V_{td}^*}{V_{cb}} - \lambda = 0. \tag{9.49}$$

Equation (9.48) requires the sum of three complex quantities to vanish. Geometrically, it defines a triangle in the (ϱ,η) plane, with sides

$$\frac{1}{\lambda}\left|\frac{V_{ub}}{V_{cb}}\right| = \sqrt{\varrho^2 + \eta^2} \equiv \mathfrak{r}_b, \quad \frac{1}{\lambda}\left|\frac{V_{td}}{V_{cb}}\right| = \sqrt{(\varrho - 1)^2 + \eta^2} \equiv \mathfrak{r}_t, \quad 1. \tag{9.50}$$

The six triangles that represent the unitarity relations (9.27) have very different shapes, yet they all contain the same area. To show this we multiply each term in (9.48) by the phase factor $V_{ud}^* V_{td}/|V_{ud}V_{td}| \equiv a/|a|$, which leaves the shape and the area of the triangle intact, with the result

$$|a| + \frac{a\,V_{us}V_{ts}^*}{|a|} + \frac{a\,V_{ub}V_{tb}^*}{|a|} = 0. \tag{9.51}$$

From (9.51) we infer that

$$\text{Area(triangle)} = \frac{1}{2}\,\text{Im}\,V_{ud}^*V_{td}V_{ts}^*V_{us} = -\frac{1}{2}\,\text{Im}\,V_{ud}^*V_{td}V_{tb}^*V_{ub}. \tag{9.52}$$

Multiplying (9.48) by $V_{us}^*V_{ts}/|V_{us}V_{ts}|$ leads to yet another expression for the same area:

$$\text{Area(triangle)} = \frac{1}{2}\,\text{Im}\,V_{us}^*V_{ts}V_{tb}^*V_{ub}. \tag{9.53}$$

Repeating this analysis for other pairs of rows or columns, we obtain the following result [59]

$$2 \times \text{Area(each triangle)} = \text{Im}\,V_{\alpha j}^*V_{\alpha k}V_{\beta k}^*V_{\beta j} \equiv J. \tag{9.54}$$

In the Wolfenstein parametrization,

$$J = A^2\lambda^6 \sin\delta. \tag{9.55}$$

All CP-violating observables within the Standard Model are proportinal to this quantity (see, for example, (9.34)).

Note that each of the subscripts in (9.54) appears twice, once with V and once with V^*. The quantity J is thus invariant under redefinitions of the phases of quark fields: $q_{\rm L} \to e^{i\phi_q}q_{\rm L}$. In other words, unlike the mixing matrix itself, J is parametrization independent.

We conclude this section with a brief review of some aspects of B^0-$\bar{B}^0$ mixing that are relevant to our discussion of the CKM parameters η and ϱ. Ignoring CP violation, the time evolution of "flavor eigenstates" $(K^0, \bar{K}^0)$ or $(B^0, \bar{B}^0)$ is given by (D.19):

$$\begin{aligned}
|M^0(t)\rangle &= f_+(t)|M^0\rangle + f_-(t)|\bar{M}^0\rangle, \\
|\bar{M}^0(t)\rangle &= f_+(t)|\bar{M}^0\rangle + f_-(t)|M^0\rangle.
\end{aligned} \tag{9.56}$$

The functions $f_+(t)$ and $f_-(t)$, defined in (D.20), can also be written as

$$\begin{aligned}
f_+(t) &= e^{-imt}e^{-\Gamma t/2}\cos\left[(\Delta m/2 - i\Delta\Gamma/4)t\right], \\
f_-(t) &= ie^{-imt}e^{-\Gamma t/2}\sin\left[(\Delta m/2 - i\Delta\Gamma/4)t\right],
\end{aligned} \tag{9.57}$$

where $\Gamma = (\Gamma_1 + \Gamma_2)/2$ is the average width of M_1^0 and M_2^0, $m = (m_1 + m_2)/2$, $\Delta m = m_2 - m_1$ and $\Delta\Gamma = \Gamma_2 - \Gamma_1$ (we define 1 and 2 such that $\Delta m > 0$).

The probabilities of finding $|M^0\rangle$ or $|\bar{M}^0\rangle$ at time t are (see also (1.23) and (1.24))

$$P(M^0 \rightarrow M^0) = P(\bar{M}^0 \rightarrow \bar{M}^0) = |f_+(t)|^2$$
$$= \frac{e^{-\Gamma t}}{2} \left[\cosh(\Delta\Gamma/2)t + \cos\Delta mt\right],$$

$$P(M^0 \rightarrow \bar{M}^0) = P(\bar{M}^0 \rightarrow M^0) = |f_-(t)|^2$$
$$= \frac{e^{-\Gamma t}}{2} \left[\cosh(\Delta\Gamma/2)t - \cos\Delta mt\right]. \tag{9.58}$$

The above expressions describe how an initially pure $|M^0\rangle$ or $|\bar{M}^0\rangle$ state evolves with time into a state of mixed flavor. As it decays, the system oscillates between M^0 and $\bar{M}^0$ with frequency Δm. This deviation from a simple exponential time evolution is an unambiguous sign of mixing. For oscillations to be detected, the system must not decay away too fast. The magnitudes of Δm and $\Delta\Gamma$ relative to Γ are therefore crucial parameters:

$$\frac{\Delta m}{\Gamma} \sim \frac{\text{lifetime}}{\text{mixing time}}. \tag{9.59}$$

Although (9.56)–(9.58) apply equally well to both $(K^0, \bar{K}^0)$ and $(B^0, \bar{B}^0)$, there are significant differences in the behavior of the two systems. Because of the light kaon mass, the dominant decay mode is $K_S^0 \rightarrow \pi\pi$ (the CP-odd kaon state decays into the phase-space suppressed 3π mode); hence $\Gamma_S \gg \Gamma_L$. In contrast, B_d^0 and B_s^0 have a number of important decay modes. However, the channels that are common to both B^0 and $\bar{B}$ (and are thus responsible for a width difference $\Delta\Gamma$) have branching ratios $\leq 10^{-3}$, leading to $\Delta\Gamma/\Gamma < 0.01$ for B_d^0 and < 0.2 for $\bar{B}_s^0$. We can therefore neglect $\Delta\Gamma/\Gamma$ in the case of B^0-$\bar{B}^0$ mixing.

On the other hand, $\Delta m(B^0) \gg \Delta m_k$ (see (9.45) and the text below). Experimentally [57, 58],

$$x_k \equiv \frac{\Delta m_k}{\Gamma_S} = 0.473 \pm 0.0018,$$

$$x_d \equiv \frac{\Delta m(B_d^0)}{\Gamma_d} = 0.728 \pm 0.025, \tag{9.60}$$

$$x_s \equiv \frac{\Delta m(B_s^0)}{\Gamma_s} > 10.5.$$

The parameter x expresses the oscillation frequency in terms of the average lifetime.

Using (9.58), we obtain the following ratio of time-integrated probabilities:

$$r \equiv \frac{\int_0^\infty P(B^0 \rightarrow \bar{B}^0)dt}{\int_0^\infty P(B^0 \rightarrow B^0)dt}$$
$$= \frac{(\Delta m)^2 + (\Delta\Gamma/2)^2}{2\Gamma^2 + (\Delta m)^2 - (\Delta\Gamma/2)^2} \approx \frac{x^2}{2 + x^2}, \quad 0 \leq r \leq 1. \tag{9.61}$$

Another useful parameter is the oscillation probability:

$$\chi \equiv \frac{\int_0^\infty \mathsf{P}(B^0 \to \bar{B}^0)dt}{\int_0^\infty \mathsf{P}(B^0 \to B^0)dt + \int_0^\infty \mathsf{P}(B^0 \to \bar{B}^0)dt}$$

$$= \frac{r}{1+r}, \quad 0 \le \chi \le \frac{1}{2}. \tag{9.62}$$

The measurement of B^0-$\bar{B}^0$ mixing requires the flavour quantum number, B, of the neutral meson to be identified at both its production and decay. Since B (like S) is conserved in strong and electromagnetic interactions, B mesons are produced in pairs. The flavour of B^0 ($\bar{B}^0$) can be traced by observing semileptonic decays $B^0 \to \ell^- \nu X$ and $\bar{B}^0 \to \ell^+ \bar{\nu} X$. One thus expects

$$B^0 \overset{\text{mix}}{\longrightarrow} \bar{B}^0 \overset{\text{decay}}{\longrightarrow} \ell^+ \bar{\nu} X. \tag{9.63}$$

Experimentally, the amount of mixing is determined through

$$\mathcal{R} = \frac{N(BB) + N(\bar{B}\bar{B})}{N(B\bar{B}) + N(\bar{B}B)} = \frac{N(\ell^- \ell^-) + N(\ell^+ \ell^+)}{N(\ell^+ \ell^-)}, \tag{9.64}$$

where $N(BB)$ is the number of BB final states in a sample of events from a process where a $B\bar{B}$ pair is initially produced, etc. Note that $N(B\bar{B})$ and $N(\bar{B}B)$ are experimentally indistinguishable.

When a $B\bar{B}$ pair is produced incoherently,[36] which occurs at energies well above the $b\bar{b}$ threshold, the time evolution of one meson is independent of the other. In this case,

$$\mathcal{R} = \frac{2\chi(1-\chi)}{(1-\chi)^2 + \chi^2} = \frac{2r}{1+r^2}, \quad \text{incoherent production}, \tag{9.65}$$

since $\mathsf{P}(BB) = \mathsf{P}(\bar{B}\bar{B}) = \mathsf{P}(B_a \text{ oscillates}) \times \mathsf{P}(B_b \text{ remains the same}) = \chi(1-\chi)$, $\mathsf{P}(B\bar{B}) = (1-\chi)^2$ (neither oscillates) and $\mathsf{P}(\bar{B}B) = \chi^2$ (both oscillate). At LEP and hadron colliders, where both the B_d^0 and B_s^0 are produced, one measures the sum of their mixing probabilities, weighted by the corresponding production fractions: $\bar{\chi} = f_d \chi_d + f_s \chi_s$.

The situation is quite different on the $\Upsilon(4S)$ resonance, or at the $B\bar{B}^*$ threshold, where the two mesons are produced coherently (i.e., they form a quantum-mechanical state of definite orbital angular momentum, ℓ, and parity). The $\Upsilon(4S)$ resonance is a P wave $b\bar{b}$ bound state with $C = -1$ and $P = -1$ that lies just above the $B_d \bar{B}_d$ threshold (its mass is less than $2\,m(B_s^0)$). The $\Upsilon(4S)$ state decays strongly into $B^+ B^-$ or $B^0 \bar{B}^0$ (see (9.41)). Since it is produced via a "virtual" photon, the $B^0 \bar{B}^0$ pair is in a pure $C = -1$ quantum state (see Sects. 1.4 and 7.1):

$$|B^0 \bar{B}^0\rangle = \frac{1}{\sqrt{2}} \left\{ |B^0(\boldsymbol{k})\rangle |\bar{B}^0(-\boldsymbol{k})\rangle - |\bar{B}^0(\boldsymbol{k})\rangle |B^0(-\boldsymbol{k})\rangle \right\}. \tag{9.66}$$

The two B mesons are strongly correlated: at no time can the original $B^0 \bar{B}^0$ system evolve into two identical states in the $\Upsilon(4S)$ rest frame. As

---

[36] That is, the angular momentum and parity of the pair are different for each event, i.e., the final state is a superposition of many angular-momentum states.

explained in Sect. 7.1, if the mesons were to decay at the same time to the same final state, there would be two identical $J = 0$ bosonic systems in an overall P wave. But this would violate the rule that two identical spinless bosons cannot be in an antisymmetric spatial state.

The B^0 and $\bar{B}^0$ propagate coherently until one of them decays. Only then will the state of the second particle be uniquely defined: it will have the flavor quantum number opposite to that of the first B meson.

Suppose that the two decays occur at times t_1 and t_2. Using expressions (9.56), which describe the time evolution of flavor eigenstates $|B^0\rangle$ and $|\bar{B}^0\rangle$, we obtain (with $f_\pm^1 \equiv f_+(t_1)$, etc.)

$$|B^0\bar{B}^0(t)\rangle \propto \left(f_+^1 f_-^2 - f_-^1 f_+^2\right)|B^0 B^0\rangle + \left(f_+^1 f_+^2 - f_-^1 f_-^2\right)|B^0\bar{B}^0\rangle$$
$$+\left(f_-^1 f_-^2 - f_+^1 f_+^2\right)|\bar{B}^0 B^0\rangle + \left(f_-^1 f_-^2 - f_+^1 f_+^2\right)|\bar{B}^0\bar{B}^0\rangle. \quad (9.67)$$

From (9.67) we see that $\mathrm{P}(B^0 B^0) = \mathrm{P}(\bar{B}^0\bar{B}^0)$ and $\mathrm{P}(B^0\bar{B}^0) = \mathrm{P}(\bar{B}^0 B^0)$; hence,

$$\mathcal{R} = \frac{\int_0^\infty |f_+(t_1)f_-(t_2) - f_-(t_1)f_+(t_2)|^2\, dt_1 dt_2}{\int_0^\infty |f_+(t_1)f_+(t_2) - f_-(t_1)f_-(t_2)|^2\, dt_1 dt_2} \equiv \frac{\mathcal{N}}{\mathcal{D}}. \quad (9.68)$$

Writing

$$f_\pm(t) = \frac{1}{2}e^{-imt}e^{-\Gamma t/2}\left[e^{i\Delta mt/2}e^{\Delta\Gamma t/4} \pm e^{-i\Delta mt/2}e^{-\Delta\Gamma t/4}\right], \quad (9.69)$$

it follows that

$$\left.\begin{array}{c}\mathcal{N}\\\mathcal{D}\end{array}\right\} = \int_0^\infty dt_1 dt_2 e^{-\Gamma t}\left[e^{\Delta\Gamma\Delta t/2} + e^{-\Delta\Gamma\Delta t/2} \mp 2\mathrm{Re}\, e^{-i\Delta m\Delta t}\right]$$

$$= \int_0^\infty dt_1 dt_2\left[e^{-\Gamma_2 t_1}e^{-\Gamma_1 t_2} + e^{-\Gamma_1 t_1}e^{-\Gamma_2 t_2}\right.$$

$$\left.\mp 2\mathrm{Re}\, e^{-\Gamma t_1}e^{i\Delta mt_1}e^{-\Gamma t_2}e^{-i\Delta mt_2}\right]$$

$$= \frac{2}{\Gamma_1\Gamma_2} \mp 2\mathrm{Re}\, \frac{1}{\Gamma - i\Delta m}\frac{1}{\Gamma + i\Delta m}$$

$$= \frac{2}{\Gamma^2 - (\Delta\Gamma/2)^2} \mp \frac{2}{\Gamma^2 + (\Delta m)^2}. \quad (9.70)$$

Therefore

$$\mathcal{R} = r = \frac{\chi}{1 - \chi}, \quad \text{coherent production } (\ell \text{ odd}). \quad (9.71)$$

The $\Upsilon(4S)$ resonance decays to $B^0\bar{B}^0$ or B^+B^-, and so the observed number of $N(\ell^+\ell^-)$ events has to be corrected for leptons coming from charged B mesons, a procedure that is not entirely unambiguous.

The first evidence for $B\bar{B}$ mixing was provided in 1987 by the UA1 experiment[37] at the CERN $p\bar{p}$ collider [60]. Soon thereafter the ARGUS collaboration at the DORIS e^+e^- storage ring observed large B_d^0-$\bar{B}_d^0$ mixing

---

[37] In 1983, the UA1 collaboration, led by C. Rubbia, discovered the intermediate vector bosons $W^\pm$ and Z^0.

($r = 0.21 \pm 0.08$) among B mesons produced in $\Upsilon(4S)$ decays [61]. Their result strongly suggested that the top quark was much heavier than expected.

Until recently, all measurements of $B\bar{B}$ mixing were time integrated. These studies are insensitive to x when mixing is maximal because $x \to \infty$ as $\chi \to 0.5$ (see (9.60)–(9.62)). To measure B_s^0-$\bar{B}_s^0$ transitions one thus needs to determine the time evolution of B_s^0 mesons, which is not an easy task given their rapid oscillation rate.

The oscillation period gives a direct measurement of the mass difference between the CP eigenstates B_1^0 and B_2^0, provided the proper time of the B-meson decay, t_p, is known with sufficient accuracy: $t_p = L/\beta\gamma = L(m/p)$, where L is the measured decay length; m and p are are the mass and momentum of the meson, respectively. The typical experimental resolution of LEP experiments is 2.5 ps in t_p.

The value of Δm is found from the fraction of "mixed" or "unmixed" events as a function of t_p by using (9.58). Based on data collected between 1991 and 1994, the DELPHI collaboration at LEP has reported [58]

$$\Delta m(B_s^0) > 6.5\,\text{ps}^{-1} \ (4.3 \times 10^{-3}\,\text{eV}) \quad \text{at 95\% CL} \tag{9.72}$$

corresponding to $x_s > 10.5$, where $x_s = \Delta m(B_s^0)\tau(B_s^0)$ and $\tau(B_s^0) = (1.61 \pm 0.1)\,\text{ps}$.

As we mentioned earlier, the measurement of $|V_{td}|$ suffers from large theoretical uncertainties associated with $f_B\sqrt{\mathcal{B}_B}$. This uncertainty can be reduced by measuring

$$\frac{\Delta m(B_s^0)}{\Delta m(B_d^0)} = \frac{m_{B_s}}{m_{B_d}} \left|\frac{V_{ts}}{V_{td}}\right|^2 \xi_s^2 = \frac{m_{B_s}}{m_{B_d}} \frac{\xi_s^2}{\lambda^2\left[(\varrho - 1)^2 + \eta^2\right]}, \tag{9.73}$$

where ξ_s is the ratio of hadronic matrix elements for the B_s^0 and B_d^0 [54]:

$$\xi_s \equiv \frac{f_{B_s}\sqrt{\mathcal{B}_{B_s}}}{f_{B_d}\sqrt{\mathcal{B}_{B_d}}} = 1.16 \pm 0.05. \tag{9.74}$$

Unfortunately, it is much more difficult to determine $\Delta m(B_s^0)$ than $\Delta m(B_d^0)$ because (a) the fraction of B_s^0 mesons produced in b decays is considerably smaller than that of B_d^0 particles, and (b) the large value of $\Delta m(B_s^0)$ (experimentally, $\Delta m(B_s^0)/\Delta m(B_d^0) > 14$) leads to rapid oscillations that complicate the measurement.

9.3 Rare Kaon Decays

Over the past forty years, studies of rare meson decays have contributed significantly to our present understanding of weak interactions. As explained in Chap. 1, the observation of both $K \to 2\pi$ and $K \to 3\pi$ decays led to the discovery of parity violation. Parity is maximally violated in weak interactions in the sense that all neutrinos are left-handed and all antineutrinos are right-handed. This motivated Feynman, Gell-Mann and others to formulate the weak interaction in terms of vector–axial vector (V–A) currents. The helicity suppressed decay[38] $\pi \to e\nu_e$ provided crucial support for the "V–A theory",[39] which has been very successful in explaining most of the low-energy weak-interaction phenomena.

The observed violation of CP symmetry in K_L^0 decays (at a rate of about 10^{-3}) may be a fundamental property of nature, with important implications for the early evolution of the universe. A deeper insight into CP violation is expected to be gained from precision measurements of theoretically "clean" rare kaon decays, such as $K_L^0 \to \pi^0 \nu\bar{\nu}$. The suppression of the $K_L^0 \to \mu^+\mu^-$ decay (at the level of 10^{-8}), discussed in Sect. 9.1, suggested the existence of the charm quark, and thus played an important role in the development of the Standard Model (the GIM mechanism). In the same section we described how the sensitivity of K^0-$\bar{K}^0$ mixing to energies higher than the kaon mass scale was used to predict the mass of the charm quark. Similarly, rare kaon decays that are dominated by one-loop Feynman diagrams with top quark exchange can yield valuable measurements of the CKM matrix elements V_{td} and V_{ts}. Since the branching ratio for $t \to d + W$ is very small, it is difficult to determine the coupling V_{td} directly from t decays.

Rare kaon decays are an important source of information on higher-order effects in electroweak interactions, and can therefore serve as a probe into physics beyond the Standard Model. Experiments at the highest-energy particle colliders, and those studying the rarest of K-meson decays at low energies, are pursuing different aspects of the same physics.

In what follows we will concentrate on those processes that are theoretically best understood. Our exposition is meant to be pedagogical, rather than comprehensive, in order to highlight the underlying physics. We will not discuss decays that violate lepton number conservation.

9.3.1 $K_L^0 \to \pi^0 \nu\bar{\nu}$ and $K^+ \to \pi^+ \nu\bar{\nu}$

Within the Standard Model, these transitions are loop-induced semileptonic decays of the type $s \to d + \ell + \bar{\ell}$. They are entirely due to second-order weak processes determined by Z^0-penguin and W-box diagrams: since photons do not couple to neutrinos, there is no electromagnetic contribution.

---

[38] Since the pion has spin zero and, according to the V–A law, the neutrino is left-handed, the lepton in $\pi \to \ell\nu_\ell$ must have negative helicity ($\lambda = -1$). The probability for a lepton of velocity v to be left-handed is $P(\lambda = -1) = 1 - v/c$. This probability is much smaller for the light electron than for the muon ($m_\mu/m_e \approx 200$). The electronic decay mode is even more suppressed in the $K \to \ell\nu_\ell$ case because the electron is more relativistic than in the pion decay. The phase space can do little to improve its odds against the muonic decay mode.

[39] In fact, the suppression is proportional to $(m_e/m_\mu)^2$ for any arbitrary mixing of V and A couplings: $\mathcal{J}_\alpha^\ell = \bar{u}_\ell \gamma_\alpha (C_V + C_A \gamma_5) v_\nu \to \Gamma(\pi \to \ell\nu_\ell) \propto 4(C_V^2 + C_A^2)m_\ell^2$.

Both decays are theoretically "clean" because the hadronic transition amplitudes are matrix elements of quark currents between mesonic states, which can be extracted from the leading semileptonic decays by using isospin symmetry.

The process $K_L^0 \to \pi^0 \nu \bar{\nu}$ offers the most transparent window into the origin of CP violation proposed so far. It proceeds almost entirely through direct CP violation [62], and is completely dominated by "short-distance" loop diagrams with top quark exchange. Although this decay has a miniscule branching ratio (about 10^{-11}) and is experimentally very challenging, its measurement, which is complementary to those planned in the B^0 system, is feasible and certainly worth the effort.

The main features of the decay $K \to \pi \nu \bar{\nu}$, summarized above, can be discerned from the relatively simple calculation of the box diagram in Fig. 9.6a. Neglecting the charged lepton mass and external quark momenta, the corresponding amplitude reads (the contribution of unphysical scalars in the t'Hooft–Feynman gauge can be ignored because their coupling to leptons is proportional to $m_\ell/m_w \ll 1$)

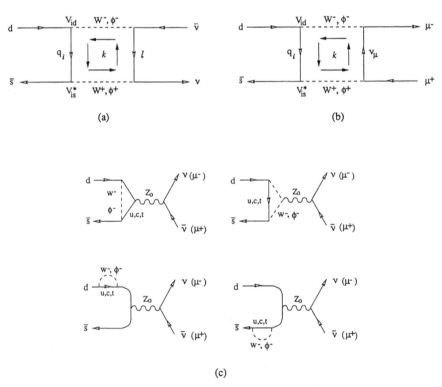

(a)   (b)

(c)

Fig. 9.6. Electroweak diagrams for $K \to \pi \nu \bar{\nu}$ (**a,c**) and $K_L^0 \to \mu^+ \mu^-$ (**b,c**)

$$i\mathcal{M} = \sum_{i=u,c,t} V_{id}V_{is}^* \int \frac{\mathrm{d}^4 k}{(2\pi)^4}\, \bar{v}_s \left[\frac{-ig}{\sqrt{2}}\gamma^\mu L\right] \frac{i(\hat{k}+m_i)}{k^2-m_i^2}\left[\frac{-ig}{\sqrt{2}}\gamma^\nu L\right]$$

$$\times u_d \left(\frac{-ig_{\nu\sigma}}{k^2-m_w^2}\right)\left(\frac{-ig_{\mu\varrho}}{k^2-m_w^2}\right)\bar{u}_\nu \left[\frac{-ig}{\sqrt{2}}\gamma^\varrho L\right]\frac{i}{-\hat{k}}\left[\frac{-ig}{\sqrt{2}}\gamma^\sigma L\right]v_\nu,$$

(9.75)

i.e.

$$i\mathcal{M} = \frac{-g^4}{(2\pi)^4}\sum_{i=u,c,t} V_{id}V_{is}^* \left[\bar{v}_s\gamma^\beta L u_d\right]\left[\bar{u}_\nu\gamma_\beta L v_\nu\right]$$

$$\times \int \frac{\mathrm{d}^4 k}{(k^2-m_i^2)(k^2-m_w^2)^2},$$

(9.76)

where we used $k_\beta k_\delta = g_{\beta\delta}k^2/4$ and

$$\gamma^\mu\gamma^\beta\gamma^\nu(1-\gamma_5)\otimes\gamma_\mu\gamma_\beta\gamma_\nu(1-\gamma_5) = 16\,\gamma^\alpha(1-\gamma_5)\otimes\gamma_\alpha(1-\gamma_5) \quad (9.77)$$

(note that the order of γ-matrix indices in this equation is not the same as in (9.15); hence $4 \leftrightarrow 16$). Using Feynman's parameter formula,

$$\frac{1}{a^2 b} = \int_0^1 \frac{2x\,\mathrm{d}x}{[ax+b(1-x)]^3},$$

(9.78)

the well-known expression

$$\int \frac{\mathrm{d}^4 k}{(k^2+t)^n} = i\pi^2 \frac{\Gamma(n-2)}{\Gamma(n)}\frac{1}{t^{n-2}}, \quad n \geq 3,$$

(9.79)

and

$$V_{ud}V_{us}^* = -V_{cd}V_{cs}^* - V_{td}V_{ts}^*$$

(9.80)

(see (9.28)), it is straightforward to show that

$$\mathcal{M} = \frac{-g^4}{(4\pi)^2 m_w^2}\sum_{i=c,t} V_{id}V_{is}^* \left\{\frac{x_i(x_i-\ln x_i - 1)}{(x_i-1)^2}\right\}$$

$$\times \left[\bar{v}_s\gamma^\alpha L u_d\right]\left[\bar{u}_\nu\gamma_\alpha L v_\nu\right]$$

(9.81)

with $x_i \equiv m_i^2/m_w^2$, $x_u \approx 0$, $x_c \approx 2.6\times10^{-4}$ and $x_t \approx 5$. As for (9.17), we can express this amplitude as the matrix element of an equivalent operator between the states $|K\rangle$ and $|\pi\rangle$:

$$A_\square(K \to \pi\nu\bar{\nu}) = \mathcal{C}_K \sum_{i=c,t} V_{id}V_{is}^* \mathcal{F}_\square(x_i)$$

$$\times \langle\pi\,|\,\bar{s}\gamma^\alpha d\,|\,K\rangle\left[\bar{\nu}\gamma_\alpha(1-\gamma_5)\nu\right],$$

(9.82)

where $\square$ stands for "box diagram",

$$\mathcal{C}_K \equiv \frac{-G_F}{\sqrt{2}}\frac{\alpha}{2\pi\sin^2\theta_w}, \quad \mathcal{F}_\square(x_i) = \frac{x_i(x_i-\ln x_i-1)}{(x_i-1)^2}$$

(9.83)

and θ_w is the weak mixing, or Weinberg, angle: $e = g \sin \theta_w$, $\alpha \equiv e^2/4\pi$. Since π and K have the same parity, only the vector current contributes to $\langle \pi \mid J_h^\alpha \mid K \rangle$.

The amplitudes $\langle \pi | \bar{s}\gamma^\alpha d | K \rangle$ are much simpler objects than the matrix element of the four-quark operator in (9.17). In the limit of exact isospin symmetry, which is a very good approximation, $\langle \pi^+ | \bar{s}\gamma^\alpha d | K^+ \rangle = \sqrt{2}\, \langle \pi^0 | \bar{s}\gamma^\alpha d | K^0 \rangle$. Moreover, the matrix element of the weak current $\bar{s}\gamma^\alpha d$ between K^+ and π^+ is related by isospin to the known matrix element of the operator $\bar{s}\gamma^\alpha u$ between K^+ and π^0:

$$\langle \pi^+ | \bar{s}\gamma^\alpha d | K^+ \rangle = \sqrt{2}\, \langle \pi^0 \mid \bar{s}\gamma^\alpha u \mid K^+ \rangle. \tag{9.84}$$

The operator $\bar{s}\gamma^\alpha u$ is measured in the decay $K^+ \to \pi^0 e^+ \nu_e$. The amplitude for this transition is given by (see Fig. 9.1c)

$$\mathcal{A}(K^+ \to \pi^0 e^+ \nu_e) = \frac{G_F}{\sqrt{2}} V_{us} \langle \pi^0 \mid \bar{s}\gamma^\alpha u \mid K^+ \rangle\, [\bar{\nu}\gamma_\alpha(1 - \gamma_5)e]. \tag{9.85}$$

Neglecting the positron mass, the branching ration for $K^+ \to \pi^+ \nu\bar{\nu}$ per neutrino flavor reads (the decays $K^+ \to \pi^+ \nu\bar{\nu}$ and $K^+ \to \pi^0 e^+ \nu_e$ have essentially the same phase space)

$$\mathcal{B}(K^+ \to \pi^+ \nu\bar{\nu}) = \mathcal{B}(K^+ \to \pi^0 e^+ \nu_e) \left[\frac{G_F}{\sqrt{2}} \frac{\alpha}{2\pi \sin^2 \theta_w} \mathcal{D} \Big/ \frac{G_F}{\sqrt{2}} V_{us} \right]^2 (\sqrt{2})^2$$

$$= \mathcal{B}(K^+ \to \pi^0 e^+ \nu_e) \frac{\alpha^2}{2\pi^2 \sin^4 \theta_w} \frac{|\mathcal{D}|^2}{\lambda^2}. \tag{9.86}$$

The complex coefficient $\mathcal{D}$ depends on the charm and top quark masses:

$$\mathcal{D} = \sum_{i=c,t} V_{id} V_{is}^* \, \mathcal{F}(x_i). \tag{9.87}$$

To show that the decay $K_L^0 \to \pi^0 \nu\bar{\nu}$ is CP violating, consider the behavior of the corresponding interaction lagrangian under CP:

$$\mathcal{L} = \Phi_{K_L} \overset{\leftrightarrow}{\partial^\mu} \Phi_{\pi^0} \bar{\nu}\gamma_\mu(1 - \gamma_5)\nu \xrightarrow{CP} [-\Phi_{K_L}] \overset{\leftrightarrow}{\partial_\mu} [-\Phi_{\pi^0}][-\bar{\nu}\gamma^\mu(1 - \gamma_5)\nu]$$

$$= -\mathcal{L}.$$

Using (3.2), (1.8) and (9.82), we can write [62]

$$\mathcal{A}(K_L^0 \to \pi^0 \nu\bar{\nu}) = \epsilon\, \mathcal{A}(K_1^0 \to \pi^0 \nu\bar{\nu}) + \mathcal{A}(K_2^0 \to \pi^0 \nu\bar{\nu}), \tag{9.88}$$

where[40]

---

[40] Note that $\langle \pi^0 | \bar{s}\gamma_\mu d | K^0 \rangle = \langle \pi^0 |(\hat{C}\hat{P})^\dagger \hat{C}\hat{P}\bar{s}\gamma_\mu d(\hat{C}\hat{P})^{-1}\hat{C}\hat{P}| K^0 \rangle = \langle \pi^0 | \bar{d}\gamma_\mu s | \bar{K}^0 \rangle$ (see Appendix E).

$$\mathcal{A}(K_1^0 \to \pi^0 \nu\bar{\nu}) = \frac{1}{\sqrt{2}} \left[\mathcal{A}(K^0 \to \pi^0 \nu\bar{\nu}) + \mathcal{A}(\bar{K}^0 \to \pi^0 \nu\bar{\nu}) \right]$$

$$= \mathrm{Re}\ \mathcal{A}(K^+ \to \pi^+ \nu\bar{\nu}),$$

$$\mathcal{A}(K_2^0 \to \pi^0 \nu\bar{\nu}) = \frac{1}{\sqrt{2}} \left[\mathcal{A}(K^0 \to \pi^0 \nu\bar{\nu}) - \mathcal{A}(\bar{K}^0 \to \pi^0 \nu\bar{\nu}) \right]$$

$$= i\,\mathrm{Im}\ \mathcal{A}(K^+ \to \pi^+ \nu\bar{\nu}).$$

(9.89)

Summed over three neutrino flavors, the branching ratios for the indirect and direct CP-violating contributions are, respectively,

$$\mathcal{B}(K_L^0 \to \pi^0 \nu\bar{\nu})_{\text{indirect}} \approx 3\,|\epsilon|^2 \frac{\tau_{K_L}}{\tau_{K^+}} \times 2.8 \times 10^{-6}$$

$$\times \left[\mathcal{F}(x_c) + A^2\lambda^4(1-\varrho)\mathcal{F}(x_t) \right]^2$$

$$\mathcal{B}(K_L^0 \to \pi^0 \nu\bar{\nu})_{\text{direct}} \approx 3\,\frac{\tau_{K_L}}{\tau_{K^+}} \times 2.8 \times 10^{-6} \left[A^2\lambda^4\eta\,\mathcal{F}(x_t) \right]^2$$

(9.90)

for $\mathcal{B}(K^+ \to \pi^0 e^+ \nu_e) = 0.0482$, $\sin^2\theta_{\mathrm{w}} = 0.23$ and $\alpha(m_w) = 1/128$.

The small value of ϵ renders the contribution from indirect CP violation (and hence from the charm quark) insignificant. Therefore,

$$\mathcal{B}(K_L^0 \to \pi^0 \nu\bar{\nu}) \approx \mathcal{B}(K_L^0 \to \pi^0 \nu\bar{\nu})_{\text{direct}} = 8 \times 10^{-11} \left[\eta\,\mathcal{F}(x_t) \right]^2 \quad (9.91)$$

based on $\tau_{K_L} = 4.18\tau_{K^+}$ and $A = 0.8$.

To complete the calculation of $\mathcal{B}(K \to \pi\nu\bar{\nu})$, we consider the remaining diagrams in Fig. 9.6. The presence of unphysical scalars in Fig. 9.6c cannot be ignored because of the large top quark mass. The result of a somewhat lengthy calculation yields

$$A_Z(K \to \pi\nu\bar{\nu}) = \mathcal{C}_K \sum_{i=c,t} V_{id} V_{is}^* \,\mathcal{F}_Z(x_i)\, \langle\pi|(\bar{s}d)_\mathrm{V}|K\rangle\, (\bar{\nu}\nu)_{\mathrm{V-A}}, \quad (9.92)$$

where

$$\mathcal{F}_Z(x_i) = \frac{x_i \left[x_i^2 + x_i(3\ln x_i - 7) + 2\ln x_i + 6 \right]}{8(x_i - 1)^2}. \quad (9.93)$$

Combining this result with (9.83), it follows that

$$\mathcal{F}(x_i) = \mathcal{F}_Z(x_i) + \mathcal{F}_\square(x_i) = \frac{x_i}{8} \left[\frac{3x_i - 6}{(x_i - 1)^2} \ln x_i + \frac{2 + x_i}{x_i - 1} \right]. \quad (9.94)$$

We thus finally obtain, for $\eta = 0.36$ and $m_t = 180\,\mathrm{GeV}/c^2$,

$$\mathcal{B}(K_L^0 \to \pi^0 \nu\bar{\nu}) \approx 2.8 \times 10^{-11}. \quad (9.95)$$

Isospin-violating quark mass effects and electroweak radiative corrections reduce this branching ratio by 5.6% [63]. Next-to-leading-order QCD effects are known to within ±1% [54]. The overall theoretical ambiguity in the calculation of $\mathcal{B}(K_L^0 \to \pi^0 \nu\bar{\nu})$ is below 2%. This uncertainty does not include the error on the CKM parameter η, as given by the (correlated) constraints (9.40) and (9.47).

The detection of $K_L^0 \to \pi^0 \nu \bar{\nu}$ presents a formidable challenge. The experimental signature of this decay is a single unbalanced π^0, which makes background rejection very difficult. The direction of photons from the decay $\pi^0 \to 2\gamma$ can be determined through their conversion to $e^+ e^-$ pairs. In general, the most important backgrounds are $K_L^0 \to 2\gamma$ ($\mathcal{B} \approx 5 \times 10^{-4}$), $K_L^0 \to 2\pi^0$ ($\mathcal{B} \approx 10^{-3}$), neutron interactions at residual gas atoms in the decay region that produce π^0s, $\Lambda \to n\pi^0$ decays, etc. The $K_L^0 \to 2\gamma$ decay, for example, can be discriminated by using both the transverse momentum balance of the two-gamma system and the position of the detected photons with respect to the beam axis.

Alternatively, the final state can be defined by selecting those events in which the π^0 undergoes the Dalitz decay $\pi^0 \to e^+ e^- \gamma$. In this case it is possible to reconstruct the vertex of the decay and hence the invariant mass of the π^0. Another advantage over the 2γ final state is that a relatively wide beam can be used. However, this method has the disadvantage that (a) the $\pi^0 \to e^+ e^- \gamma$ decay has a small branching ratio (about 1%) and (b) the final state in the radiative decay $K_L^0 \to \pi^\pm e^\mp \gamma \nu$ looks like $e^+ e^- \gamma$ + "nothing" if the $\pi^\pm$ is misidentified as $e^\pm$.

All attempts to detect $K_L^0 \to \pi^0 \nu \bar{\nu}$ thus far have relied on the Dalitz decay mode. The best published limit to date is $\mathcal{B}(K_L^0 \to \pi^0 \nu \bar{\nu}) < 5.8 \times 10^{-5}$ (90% CL) from Fermilab experiment E731/799 [64].

There are several proposals to measure $\mathcal{B}(K_L^0 \to \pi^0 \nu \bar{\nu})$. The KTeV experiment, described in Sect. 8.1, is expected to reach a sensitivity of 10^{-8} by identifying π^0s through the Dalitz decay. The KAMI collaboration [65] has proposed to use the Main Injector at Fermilab as a source of very high-intensity and high-energy neutral kaons, and to detect $\pi^0 \to 2\gamma$ decays in the pure CsI crystals of the KTeV apparatus. They aim at a sensitivity of better than 10^{-12}. An experiment at the KEK laboratory [66] intends to employ an array of CeF_3 crystals to measure the energy and position of the two gammas from the $\pi^0 \to \gamma\gamma$ decay, and a lead/scintillator barrel calorimeter to select two-photon events. The experiment E926 at Brookhaven [67] would exploit high beam intensities of the AGS proton synchrotron, which will be able to provide, by the year 2000, over 10^{14} protons/pulse. The Brookhaven group proposes to obtain low-momentum kaons ($\langle P_k \rangle = 700$ MeV/c) from a micro-bunched proton beam. This would allow them to determine the momentum of the decaying K_L^0 using time of flight measurements.

The expression for $\mathcal{B}(K^+ \to \pi^+ \nu \bar{\nu})$ with three quark and lepton families was originally derived by T. Inami and C. Lim [55]:

$$\mathcal{B}(K^+ \to \pi^+ \nu\bar{\nu}) = \frac{\alpha^2(m_w)\mathcal{B}(K^+ \to \pi^0 e^+ \nu_e)}{2\pi^2 \sin^4 \theta_w}$$

$$\times \sum_{\ell=e,\mu,\tau} \frac{\left|\sum_{i=c,t} V_{id}V_{is}^* \mathcal{F}(x_i, y_\ell)\right|^2}{|V_{us}|^2} \qquad (9.96)$$

(cf. (9.86) and (9.87)), with

$$\mathcal{F}(x, y) = \frac{\ln y}{16} \frac{xy}{x-y} \left(\frac{y-4}{y-1}\right)^2$$

$$+ \frac{x \ln x}{16} \left[\frac{x}{y-x}\left(\frac{x-4}{x-1}\right)^2 + 1 + \frac{3}{(x-1)^2}\right]$$

$$+ \frac{x}{8} - \frac{3x}{16(x-1)}\left(1 + \frac{3}{y-1}\right). \qquad (9.97)$$

In the above two equations, $x_i \equiv m_i^2/m_w^2$, $i = c, t$ (quarks) and $y_\ell \equiv m_\ell^2/m_w^2$, $\ell = e, \mu, \tau$ (leptons). For $y \to 0$, (9.97) reduces to (9.94). Experimentally, $x_t \approx 5$, $x_c, y_\tau \approx 10^{-4}$ ($m_\tau \approx 1.78\,\text{GeV}/c^2$), $y_\mu \approx 10^{-6}$ and $y_e \approx 10^{-11}$. We can thus write

$$\mathcal{F}(x_t, y_\ell) \approx \mathcal{F}(x_t), \quad \mathcal{F}(x_c, y_\ell) \approx x_c\left[\frac{x_c \ln x_c - y_\ell \ln y_\ell}{y_\ell - x_c} + \frac{\ln x_c - 1}{4}\right] \quad (9.98)$$

and

$$\mathcal{B}(K^+ \to \pi^+ \nu\bar{\nu}) \approx 2.8 \times 10^{-6} \left\{3\left[A^2 \lambda^4 \eta\, \mathcal{F}(x_t)\right]^2\right.$$

$$\left. + \sum_{\ell=e,\mu,\tau}\left[\mathcal{F}(x_c, y_\ell) + A^2\lambda^4(1-\varrho)\mathcal{F}(x_t)\right]^2\right\}, \quad (9.99)$$

where, to a very good approximation, $\mathcal{F}(x_c, y_e) \approx \mathcal{F}(x_c, y_\mu) \approx \mathcal{F}(x_c, 0)$. In contrast to the CP-violating decay $K_L^0 \to \pi^0 \nu\bar{\nu}$, the charm and top quark contributions to $K^+ \to \pi^+ \nu\bar{\nu}$ are of comparable size: the smallness of $\mathcal{F}(x_c, y_\ell)$ in comparison with $\mathcal{F}(x_t)$ is compensated by the strong CKM suppression of the t contribution.

Isospin-violating quark-mass effects and electroweak radiative corrections (which do not affect the short-distance structure of $K \to \pi\nu\bar{\nu}$) result in a decrease of the branching ratio by 10% [63]. Possible long-distance contributions are estimated to be negligibly small [68]. Short-distance QCD effects represent the most important class of radiative corrections to this process. The inclusion of next-to-leading-order QCD corrections yields [69, 54]

$$0.88 \times 10^{-10} \leq \mathcal{B}(K^+ \to \pi^+ \nu\bar{\nu}) \leq 1.02 \times 10^{-10}, \text{ QCD uncertainty.} (9.100)$$

for $\tilde{m}_c = 1.3\,\text{GeV}/c^2$, $\tilde{m}_t = 170\,\text{GeV}/c^2$, $\varrho = 0$ and $\eta = 0.36$. The above theoretical ambiguity is associated mainly with the charm quark. Based on our present knowledge of Standard Model parameters, Buchalla et al. predict [54]

$$0.6 \times 10^{-10} \le \mathcal{B}(K^+ \to \pi^+\nu\bar{\nu})_{\text{theor}} \le 1.5 \times 10^{-10}. \tag{9.101}$$

The determination of $|V_{td}|$ from $\mathcal{B}(K^+ \to \pi^+\nu\bar{\nu})$ and of (ϱ, η) by using both $\mathcal{B}(K^+ \to \pi^+\nu\bar{\nu})$ and $\mathcal{B}(K_L^0 \to \pi^0\nu\bar{\nu})$ is discussed in [69, 54] (see Fig. 9.7). The expected accuracy is comparable to that one can achieve by studying CP asymmetries in B decays [67].

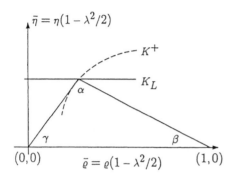

Fig. 9.7. Determination of ρ and η from $K \to \pi\nu\bar{\nu}$

The decay $K^+ \to \pi^+\nu\bar{\nu}$ occurs at a very low rate in the presence of large background sources of pions and muons, and has only the weak kinematic constraint $P_\pi \le (m_k^2 - m_\pi^2)/2m_k =227$ MeV/c. Fortunately, each of the two major K^+ decay modes $K^+ \to \pi^+\pi^0$ and $K^+ \to \mu^+\nu_\mu$ produces a single charged track of unique center-of-mass momentum (205 MeV/c and 236 MeV/c, respectively), i.e., it yields a kinematic peak that can be avoided with sufficiently good momentum resolution. Unambiguous π-μ identification and efficient photon detection are also required to reject these and other background decays. The only other important background sources are scattering of beam pions and K^+ "charge-exchange" reactions (see (1.42)) resulting in $K_L^0 \to \pi^+\ell^-\bar{\nu}$ decays, where $\ell = e$, μ is missed.

In the experiment E787 at Brookhaven [70], K^+ mesons are stopped in a target at the center of a hermetic detector (see Fig. 9.8). The momentum, kinetic energy and range of the charged decay products are measured to distinguish pions from muons. In addition, the pulse shape in a stopping counter is used to identify the decay sequence $\pi^+ \to \mu^+ \to e^+$. For their 1995 data-taking, kaons of 790 MeV/c were delivered at a rate of 7×10^6 per 1.6 s spill of the AGS proton synchrotron. The pion contamination in the beam was reduced to about 25% by using a particle separator (see Sect. 8.1). The kaons are identified by Čerenkov, tracking and energy-loss counters. Before being brought to rest in a highly segmented target composed of scintillating fibers, approximately 20% of the kaons pass through a BeO cylinder, where their momentum is reduced to about 300 MeV/c. Beam protons are completely ranged out and many of the remaining pions are absorbed or scattered in this

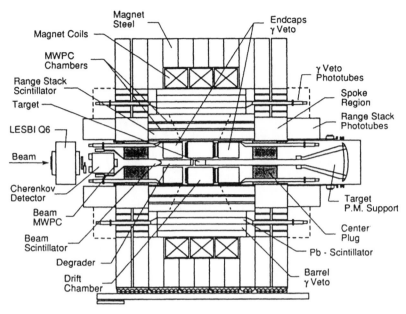

Fig. 9.8. Schematic side view of the E787 detector

"degrader". The energy and pathlength information from the target is used to correct the total energy and range determination for charged decay products. Charged particles emerging from the target are momentum analyzed as they traverse a cylindrical drift chamber. They are subsequently stopped in an array of plastic scintillation counters and straw tube tracking chambers ("range stack") to measure their range and kinetic energy. The pion detector is completely surrounded by a lead/scintillator photon detection system. The whole apparatus is contained in a solenoidal field of 1 T.

In the pion momentum region 211 MeV/c < P_π < 230 MeV/c, E787 observed one event consistent with the decay $K^+ \to \pi^+ \nu \bar{\nu}$. The background was estimated to contribute 0.08 ± 0.03 events. Assuming that the observed event is due to $K^+ \to \pi^+ \nu \bar{\nu}$, they obtained [71]

$$\mathcal{B}(K^+ \to \pi^+ \nu \bar{\nu})_{\mathrm{exp}} = 4.2^{+9.7}_{-3.5} \times 10^{-10}. \tag{9.102}$$

9.3.2 $K^0_L \to \mu^+ \mu^-$ and $K^0_L \to e^+ e^-$

The short-distance contributions to $K^0_L \to \mu^+ \mu^-$ and $K \to \pi \nu \bar{\nu}$ come from one-loop diagrams in Fig. 9.6, and are therefore very similar (the only difference is due to the reversed lepton line in the box diagram). There is another similarity between these two flavor-changing neutral-current processes: their amplitudes can be related in a simple way to the matrix elements for the charged-current transitions $K^+ \to \mu^+ \nu_\mu$ and $K^+ \to \pi^0 e^+ \nu_e$, respectively (see (9.108) and (9.84)). Neglecting the mass of the "virtual" lepton, one readily infers from the box diagrams in Figs. 9.6a,b and expressions (9.15) and (9.77):

$$\mathcal{M}_\square\left(\bar{s}d \to \begin{matrix} \mu\bar{\mu} \\ \nu\bar{\nu} \end{matrix}\right) = \begin{pmatrix} -1/4 \\ 1 \end{pmatrix} \mathcal{C}_K \sum_{i=c,t} V_{id}V_{is}^* \, \mathcal{F}_\square(x_i)(\bar{s}d)_{V-A}(\bar{\ell}\ell)_{V-A} \quad (9.103)$$

(cf. (9.81) and (9.82)), where $\ell \equiv \mu, \nu$ and $\mathcal{F}_\square(x_i)$ is given by (9.83). The penguin diagrams in Fig. 9.6c yield (cf. (9.92))

$$\mathcal{M}_Z\left(\bar{s}d \to \begin{matrix} \mu\bar{\mu} \\ \nu\bar{\nu} \end{matrix}\right) = \mathcal{C}_K \sum_{i=c,t} V_{id}V_{is}^* \, \mathcal{F}_Z(x_i)(\bar{s}d)_{V-A} \begin{pmatrix} -(\bar{\mu}\mu)_{V-A} \\ (\bar{\nu}\nu)_{V-A} \end{pmatrix} \quad (9.104)$$

with $\mathcal{F}_Z(x)$ given by (9.93). Note that the term $\sin^2\theta_{\rm w}\bar{\mu}\gamma^\alpha\mu$ is absent from the expression for $\bar{s}d \to \mu\bar{\mu}$ due to vector-current conservation (photonic penguin diagrams do not contribute to $K_{\rm L}^0 \to \mu^+\mu^-$ for the same reason). To show this, we write $\mathcal{A}(K^0 \to \mu\bar{\mu}) \propto \langle 0|\mathcal{J}_\alpha|K^0\rangle$. Since the K^0 meson is a pseudoscalar, the only kinematical quantity available to construct the hadronic matrix element is the four-momentum transfer q_α:

$$\langle 0|\mathcal{J}_\alpha|K^0\rangle \equiv f_k q_\alpha. \quad (9.105)$$

This implies that $\mathcal{J}_\alpha$ is an axial current because q_α is a polar vector (see footnote 32). Now, $\bar{\ell}q_\alpha\gamma^\alpha\ell \equiv \bar{\ell}\hat{q}\ell = \bar{\ell}(\hat{p}_\ell - \hat{p}_{\bar{\ell}})\ell = 0$, where we used the Dirac equation $(\hat{p} - m)\ell = 0$. The $\bar{s}d \to \mu\bar{\mu}$ annihilation amplitude thus reads

$$\mathcal{M}(\bar{s}d \to \mu\bar{\mu}) = -\mathcal{C}_K \sum_{i=c,t} V_{id}V_{is}^* \mathcal{G}(x_i)(\bar{s}\gamma_\alpha\gamma_5 d)(\bar{\mu}\gamma^\alpha\gamma_5\mu), \quad (9.106)$$

where

$$\mathcal{G}(x) = \frac{3}{8}\left[\frac{x}{1-x}\right]^2 \ln x + \frac{4x - x^2}{8(1-x)}. \quad (9.107)$$

We take $\mathcal{M} + \mathcal{M}^\dagger$ to be an effective hamiltonian for the decay $K_{\rm L}^0 \to \mu^+\mu^-$.[41] As a consequence of isospin invariance,

$$\langle 0 | \bar{s}\gamma_\alpha\gamma_5 u | K^+\rangle = \langle 0 | \bar{s}\gamma_\alpha\gamma_5 d | K^0\rangle \equiv f_k q_\alpha. \quad (9.108)$$

Therefore,

$$\begin{aligned} \mathcal{A}(K_{\rm L}^0 \to \mu\bar{\mu}) &= -\mathcal{C}_K \sum_{i=c,t} \mathcal{G}(x_i)\langle 0 | V_{id}V_{is}^*(\bar{s}\gamma_\alpha\gamma_5 d) \\ &\quad + V_{is}V_{id}^*(\bar{d}\gamma_\alpha\gamma_5 s)|K_{\rm L}^0\rangle(\bar{\mu}\gamma^\alpha\gamma_5\mu) \\ &= -\sqrt{2}\,\mathcal{C}_K \mathrm{Re} \sum_{i=c,t} V_{id}V_{is}^* \, \mathcal{G}(x_i)\langle 0 | \bar{s}\gamma_\alpha\gamma_5 u | K^+\rangle(\bar{\mu}\gamma^\alpha\gamma_5\mu) \\ &= \frac{\alpha\, G_{\rm F} f_k q_\alpha}{2\pi \sin^2\theta_{\rm w}} \mathrm{Re} \sum_{i=c,t} V_{id}V_{is}^* \, \mathcal{G}(x_i)(\bar{\mu}\gamma^\alpha\gamma_5\mu), \quad (9.109) \end{aligned}$$

---

[41] Since $K_{\rm L}^0$ is (mainly) CP-odd, only the CP-odd combination of axial currents $\bar{s}\gamma_\alpha\gamma_5 d + \bar{d}\gamma_\alpha\gamma_5 s$ contributes to the transition $K_{\rm L}^0 \to$ vacuum.

where we used $|K^0_L\rangle \approx [|K^0\rangle - \bar{K}^0\rangle]/\sqrt{2}$ and (see Appendix E)

$$\begin{aligned}
\langle 0\,|\bar{s}\gamma_\alpha\gamma_5 d|K^0\rangle &= \langle 0\,|\hat{C}\hat{P})^\dagger\hat{C}\hat{P}(\bar{s}\gamma_\alpha\gamma_5 d)(\hat{C}\hat{P})^{-1}\hat{C}\hat{P}|K^0(E,\mathbf{p}\,)\rangle \\
&= -\langle 0\,|\bar{d}\gamma^\alpha\gamma_5 s|\bar{K}^0(E,-\mathbf{p})\rangle,
\end{aligned}$$

i.e.,

$$\langle 0\,|\bar{s}\gamma_\alpha\gamma_5 d|K^0(E,\mathbf{p}\,)\rangle = -\langle 0\,|\bar{d}\gamma_\alpha\gamma_5 s|\bar{K}^0(E,\mathbf{p}\,)\rangle. \tag{9.110}$$

Expression (9.109) can be written as

$$\begin{aligned}
\mathcal{A}(K^0_L \to \mu\bar{\mu}) &= \frac{\alpha\,G_F f_k \tilde{D}}{2\pi\sin^2\theta_w}\,(k-p)_\alpha\bar{u}(k)\gamma^\alpha\gamma_5 v(p) \\
&= \frac{\alpha\,G_F f_k \tilde{D}}{\pi\sin^2\theta_w}\,m_\mu\bar{u}(k)\gamma_5 v(p) \tag{9.111}
\end{aligned}$$

with $q_\alpha = (k-p)_\alpha$ and

$$\tilde{D} \equiv \mathrm{Re}\sum_{i=c,t} V_{id}V^*_{is}\,\mathcal{G}(x_i). \tag{9.112}$$

Hence,

$$|\mathcal{A}|^2 = 4(G_F f_k)^2\left(\frac{\alpha\tilde{D}}{\pi\sin^2\theta_w}\right)^2 m^2_\mu\,(kp + m^2_\mu). \tag{9.113}$$

In the center-of-mass system, the decay rate is given by

$$d\Gamma = \frac{(2\pi)^4}{2m_k}\,|\mathcal{A}|^2\,\frac{d^3\mathbf{p}}{(2\pi)^3 2E_\mu}\,\frac{d^3\mathbf{k}}{(2\pi)^3 2E_\mu}\,\delta(m_k - 2E_\mu)\,\delta^{(3)}(\mathbf{k}+\mathbf{p}). \tag{9.114}$$

Using $\mathbf{p} = -\mathbf{k} \to kp = E^2_\mu + \mathbf{p}^2$ and

$$\int d^3\mathbf{k} = \int |\mathbf{k}|^2 d|\mathbf{k}|d\Omega = 4\pi\int\sqrt{E^2_\mu - m^2_\mu}\,E_\mu dE_\mu \tag{9.115}$$

it can easily be shown that

$$\Gamma(K^0_L \to \mu\bar{\mu}) = \frac{(G_F f_k)^2 m_k}{8\pi}\left(\frac{\alpha\tilde{D}}{\pi\sin^2\theta_w}\right)^2 m^2_\mu\left(1 - \frac{4m^2_\mu}{m^2_k}\right)^{1/2}. \tag{9.116}$$

The $K^+ \to \mu^+\nu_\mu$ amplitude reads (see Fig. 9.1a)

$$\begin{aligned}
\mathcal{A}(K^+ \to \mu^+\nu_\mu) &= \frac{G_F f_k}{\sqrt{2}}\,V_{us}(k-p)_\alpha\bar{u}_\nu(k)\gamma^\alpha(1-\gamma_5)v_\mu(p) \\
&= \frac{-G_F f_k}{\sqrt{2}}\,V_{us}m_\mu\bar{u}(k)(1+\gamma_5)v(p) \tag{9.117}
\end{aligned}$$

since $\bar{u}(k)\hat{k} = 0$ and $(\hat{p} - m_\mu)v(p) = 0$. Therefore,

$$|\mathcal{A}|^2 = 4(G_F f_k)^2|V_{us}|^2 m^2_\mu\,(kp). \tag{9.118}$$

If we note that $kp = E_\nu(E_\mu + E_\nu)$ and $\int d^3\mathbf{k} = 4\pi \int E_\nu^2 dE_\nu$, it follows readily from (9.114) and (9.118) that

$$\Gamma(K^+ \to \mu^+\nu_\mu) = \frac{(G_F f_k)^2 |V_{us}|^2 m_k}{8\pi} m_\mu^2 \left(1 - \frac{m_\mu^2}{m_k^2}\right)^2 \tag{9.119}$$

The branching ratio for $K_L^0 \to \mu^+\mu^-$ can thus be expressed as [55]

$$\mathcal{B}(K_L^0 \to \mu^+\mu^-) = \left(\frac{\alpha}{\pi \sin^2\theta_w}\right)^2 \frac{\tilde{D}^2}{|V_{us}|^2} \frac{(1 - 4m_\mu^2/m_k^2)^{1/2}}{(1 - m_\mu^2/m_k^2)^2} \frac{\tau_{K_L}}{\tau_{K^+}}$$
$$\times \mathcal{B}(K^+ \to \mu^+\nu_\mu). \tag{9.120}$$

The Standard Model expectation for the short-distance contribution to the $K_L^0 \to \mu^+\mu^-$ branching ratio is [54]

$$0.6 \times 10^{-9} \leq \mathcal{B}(K_L^0 \to \mu^+\mu^-)_{SD} \leq 2.0 \times 10^{-9}. \tag{9.121}$$

In passing we note that the K decay constant f_k is obtained from $K \to \mu\nu$ using

$$\Gamma(K \to \mu\nu_\mu) = \frac{(G_F f_k)^2 |V_{us}|^2 m_k}{8\pi} m_\mu^2 \left(1 - \frac{m_\mu^2}{m_k^2}\right)^2$$
$$\times \left(1 + \frac{2\alpha}{\pi}\ln\frac{m_z}{m_k} - \frac{3\alpha}{\pi}\ln\frac{m_k}{m_\mu}\right), \tag{9.122}$$

which includes the leading radiative corrections [72]. Experimentally,

$$f_k^2 |V_{us}|^2 = 1.23 \times 10^{-3}\,\text{GeV} \quad \text{or} \quad f_k = 159 \pm 1.6\,\text{MeV}. \tag{9.123}$$

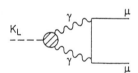

K_L

Fig. 9.9. The two-photon absorptive contribution to $K_L^0 \to \mu^+\mu^-$

It turns out that the most important contribution to $K_L^0 \to \mu^+\mu^-$ comes from the real two-photon intermediate state shown in Fig. 9.9. The corresponding absorptive (imaginary) part of the amplitude $\mathcal{A}(K_L^0 \to \mu^+\mu^-)_{2\gamma}$ can be estimated reliably: it is simply the product of the experimentally known decay amplitude $\mathcal{A}(K_L^0 \to \gamma\gamma)$ and the pure QED scattering amplitude $\mathcal{A}(\gamma\gamma \to \mu^+\mu^-)$. The decay rate for this process is given by [73][42]

---

[42] The factor $(m_\ell/m_k)^2$ in (9.124) suppresses $K_L^0 \to e^+e^-$ relative to $K_L^0 \to \mu^+\mu^-$. This is the familiar helicity suppression of the decay of a pseudoscalar meson into a fermion–antifermion pair through the axial vector coupling.

$$\frac{\Gamma(K_L^0 \to \mu^+\mu^-)_{2\gamma}}{\Gamma(K_L^0 \to \gamma\gamma)} = \frac{\alpha^2}{2} \left(\frac{m_\mu}{m_k}\right)^2 \frac{1}{\beta} \left[\ln\left(\frac{1+\beta}{1-\beta}\right)\right]^2 = 1.2 \times 10^{-5}, \quad (9.124)$$

where $\beta = (1 - 4m_\mu^2/m_k^2)^{1/2}$. Based on the experimental value [21]

$$\mathcal{B}(K_L^0 \to \gamma\gamma)_{\text{exp}} = (5.92 \pm 0.15) \times 10^{-4}, \quad (9.125)$$

this leads to

$$\mathcal{B}(K_L^0 \to \mu^+\mu^-)_{\text{abs}} = (7.1 \pm 0.18) \times 10^{-9}, \quad (9.126)$$

which pretty much accounts for the observed decay rate [21]

$$\mathcal{B}(K_L^0 \to \mu^+\mu^-)_{exp} = (7.2 \pm 0.5) \times 10^{-9}. \quad (9.127)$$

The dispersive (real) part of the amplitude $\mathcal{A}(K_L^0 \to \mu^+\mu^-)$ is the sum of long- and short-distance contributions. The long-distance part is mainly due to the virtual two-photon intermediate state ($K_L^0 \to \gamma^*\gamma^* \to \mu^+\mu^-$), and is difficult to calculate reliably (see, e.g., [107]). This is unfortunate because, as we have seen, the short-distance part is known rather well.

The experiment E871 at Brookhaven has detected approximately 6400 $K_L^0 \to \mu^+\mu^-$ events [74] (previous experiments had accumulated about 900 $K_L^0 \to \mu^+\mu^-$ decays [75]). As shown in Fig. 9.10, protons of energy 24 GeV from the AGS produce a neutral beam, in which about 2×10^7 K_L^0 mesons decay per beam spill within the vacuum decay region of the E871 detector. The neutral beam is stopped inside the spectrometer by a hadronic "beam plug". Tracking is provided by straw chambers in the upstream section of the spectrometer (where rates are the highest), and by drift chambers in the section downstream of the beam plug. The beam volume within the spectrometer is filled with helium to reduce multiple scattering and particle interactions. Trajectories of two-body charged decay products are nearly parallel to the initial beam direction downstream of two dipole magnets. A threshold Čerenkov counter and a lead-glass array are used for electron identification (the experiment also intends to search for the lepton-number-violating decay $K_L^0 \to \mu e$ at the 10^{-12} level). Muons are identified and stopped in a segmented absorber stack containing proportional wire chambers and scintillator hodoscopes. The experiment is expected to measure the $K_L^0 \to \mu^+\mu^-$ branching ratio much more precisely than its predecessor, the E791.

It remains to be seen, however, if the anticipated improvement in experimental precision can be matched by an improvement in the theoretical estimate of the long-distance dispersive contribution to this decay. Should E871 confirm that the measured branching ratio is indeed saturated by $\mathcal{B}(K_L^0 \to \mu^+\mu^-)_{\text{abs}}$, it would be difficult to understand why the two apparently unrelated dispersive contributions cancel each other.

Since the kaon is spinless, the muon pair in the decay $K^0 \to \mu^+\mu^-$ has the total angular momentum $J = 0$. This implies that $s = \ell$, where s and ℓ are the spin and orbital angular momentum of the muonic system, respectively. The replacement $\mu^+ \leftrightarrow \mu^-$ means not only an interchange of spins, but also

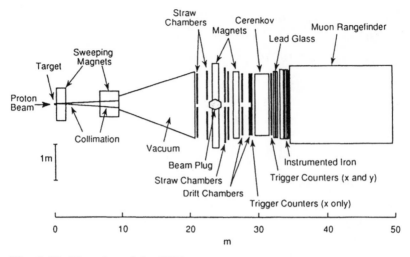

Fig. 9.10. Plan view of the E871 spectrometer

space inversion $\mathbf{r} \rightarrow -\mathbf{r}$. The combined operation of charge conjugation and parity transformation results in

$$\hat{C}\hat{P}\psi_{\mu\mu} = (-1)^{\mathrm{S}+1}\psi_{\mu\mu}. \tag{9.128}$$

Thus there are two possible states for this system, which are eigenstates of $\hat{C}\hat{P}$:

$$\begin{aligned}
\hat{C}\hat{P}|(\mu^+\mu^-)_{^3P_0}\rangle &= +|(\mu^+\mu^-)_{^3P_0}\rangle, \\
\hat{C}\hat{P}|(\mu^+\mu^-)_{^1S_0}\rangle &= -|(\mu^+\mu^-)_{^1S_0}\rangle.
\end{aligned} \tag{9.129}$$

The most general amplitude for $K_2^0 \rightarrow \mu^+\mu^-$ has the following form (cf. (9.111))

$$\mathcal{M}(K_2^0 \rightarrow \mu^+\mu^-) = \bar{u}(p_-, s_-)(a\gamma_5 + ib)v(p_+, s_+). \tag{9.130}$$

The amplitude a is CP conserving (the μ pair is in the $^1S_0$ state), whereas b is CP violating ($\mu^+\mu^-$ is in the $^3P_0$ state). The interference between the two amplitudes results in longitudinal polarization of the leptons, a CP violating observable. If we sum over the possible spin states of both muons,

$$\begin{aligned}
|\mathcal{M}|^2 &= \mathrm{Tr}\left[(\hat{p}_- + m_\mu)(a\gamma_5 + ib)(\hat{p}_- - m_\mu)(-a^*\gamma_5 - ib^*)\right] \\
&= 4\left\{p_+ \cdot p_-(|a|^2 + |b|^2) + m_\mu^2(|a|^2 - |b|^2)\right\}.
\end{aligned} \tag{9.131}$$

Using (9.114) and (9.115), it can easily be shown that the total $K_2^0 \rightarrow \mu^+\mu^-$ decay rate is given by

$$\Gamma = \frac{m_k\beta}{8\pi}\left(|a|^2 + \beta^2|b|^2\right), \tag{9.132}$$

where, as before, $\beta = (1 - 4m_\mu^2/m_k^2)^{1/2}$.

Since we are interested in polarized muons, we must insert a spin projection operator $(1+\gamma_5\hat{s})/2$ in front of the muon spinor. In the muon rest frame, where $s = (0, \boldsymbol{s})$, the operator projects onto a spin direction parallel to the unit vector $\boldsymbol{s}$. After the insertion, the probability for the decay into polarized muons can be calculated using the standard trace techniques. Summing, for example, over the spin states of the μ^+, we obtain[43]

$$
\begin{aligned}
|\mathcal{M}|^2 &= \mathrm{Tr}\left[(\hat{p}_- + m_\mu)\left(\frac{1+\gamma_5\hat{s}_-}{2}\right)(a\gamma_5 + ib)(\hat{p}_- - m_\mu)(-a^*\gamma_5 - ib^*)\right] \\
&= 2\left\{p_+ \cdot p_-(|a|^2 + |b|^2) + m_\mu^2(|a|^2 - |b|^2)\right. \\
&\quad \left. + im_\mu ab^*(p_+ - p_-)s_- - im_\mu ba^*(p_+ + p_-)s_-\right\}.
\end{aligned}
\tag{9.133}
$$

The spin 4-vector s_- can be expressed in terms of the unit vector $\boldsymbol{s}_-$ in the muon rest frame by means of the Lorentz transformation

$$
s = \left[\frac{\mathbf{p}\cdot\boldsymbol{s}}{m_\mu}, \boldsymbol{s} + \frac{(\mathbf{p}\cdot\boldsymbol{s})\mathbf{p}}{m_\mu(E_\mu + m_\mu)}\right].
\tag{9.134}
$$

By virtue of $p_+ - p_- = (0, 2\mathbf{p}_-)$ and $p_+ + p_- = (2E, 0)$, expression (9.133) becomes

$$
\begin{aligned}
|\mathcal{M}|^2 &= 2\{p_+ \cdot p_-(|a|^2 + |b|^2) + m_\mu^2(|a|^2 - |b|^2) \\
&\quad + 4E\,\mathrm{Im}\,(ba^*)(\mathbf{p}_- \cdot \boldsymbol{s}_-)\}.
\end{aligned}
\tag{9.135}
$$

The corresponding decay rate in the K_2^0 rest frame reads [76]

$$
\mathrm{d}W = \frac{m_k\beta}{64\pi^2}\left(|a|^2 + \beta^2|b|^2\right)(1 + \mathcal{P}_\mathrm{L}\boldsymbol{n}\cdot\boldsymbol{s})\mathrm{d}\Omega,
\tag{9.136}
$$

where $\boldsymbol{n} \equiv \mathbf{p}/|\mathbf{p}|$, $\boldsymbol{n}\cdot\boldsymbol{s} = \pm|\boldsymbol{n}|$ and

$$
\mathcal{P}_\mathrm{L}(K_2^0 \to \mu^+\mu^-) \equiv \frac{2\beta\,\mathrm{Im}\,(ba^*)}{|a|^2 + \beta^2|b|^2} = \frac{m_k\beta^2\,\mathrm{Im}\,(ba^*)}{4\pi\Gamma}
\tag{9.137}
$$

$\mathcal{P}_\mathrm{L}$ is the degree of longitudinal polarization of the muon:

$$
\mathcal{P}_\mathrm{L} = \frac{N_R - N_\mathrm{L}}{N_R + N_\mathrm{L}},
\tag{9.138}
$$

where $N_R(N_\mathrm{L})$ is the number of μ^-s emerging with positive (negative) helicity. The longitudinal polarization violates CP because $\mathbf{p}\cdot\boldsymbol{s}$ is invariant under charge conjugation, but changes sign under parity transformation.

We now take into account that the K_L^0 state has a small CP-even admixture: $K_\mathrm{L}^0 \approx K_2^0 + \epsilon K_1^0$. The amplitude for $K_\mathrm{L}^0 \to \mu^+\mu^-$ is given by (9.130), with

$$
a = a_2 + \epsilon a_1, \quad b = b_2 + \epsilon b_1.
\tag{9.139}
$$

---

[43] To avoid depolarization effects associated with the atomic capture of the μ^-, it is better to measure the longitudinal polarization of the μ^+.

To first order in CP-violating quantities,

$$\text{Im}\,(ba^*) \approx \text{Im}\,[a_2^*(b_2 + \epsilon b_1)]. \tag{9.140}$$

The amplitudes a_2 and b_1 are CP conserving, whereas b_2 is CP violating. In the Standard Model, b_2 is due to the $\bar{s}d \to \mu\bar{\mu}$ transition mediated by the Higgs scalar H^0. This contribution is negligible ($\mathcal{P}_{\text{L}}(\text{Higgs}) < 10^{-4}$ for $m_{\text{H}} > 10\,\text{GeV}/c^2$) in view of the LEP result $m_{\text{H}} > 58.4\,GeV/c^2$ [21].

Writing

$$\epsilon = |\epsilon|\,\text{e}^{\text{i}\phi_{\pi\pi}} \approx |\epsilon|\,\text{e}^{\text{i}\pi/4} = \frac{|\epsilon|}{\sqrt{2}}(1+\text{i}), \tag{9.141}$$

expression (9.137) and the second term in (9.140) thus give

$$\mathcal{P}_{\text{L}}(K_{\text{L}}^0 \to \mu^+\mu^-) \approx \frac{\sqrt{2}\,\beta|\epsilon|}{|a_2|^2}\{\text{Re}\,a_2(\text{Re}\,b_1 \\ + \text{Im}\,b_1) + \text{Im}\,a_2(\text{Im}\,b_1 - \text{Re}\,b_1)\}. \tag{9.142}$$

Recall that $\text{Im}\,a_2$ is completely dominated by the real two-photon intermediate state:

$$|\text{Im}\,a_2|_{2\gamma} = \frac{\alpha m_\mu}{4\beta m_k}\,\ln\left(\frac{1+\beta}{1-\beta}\right)\left[\frac{64\pi\,\Gamma(K_{\text{L}}^0 \to \gamma\gamma)}{m_k}\right]^{1/2}. \tag{9.143}$$

An upper bound on $|\text{Re}\,a_2|$ follows from (9.126) and (9.127). Based on a leading-order calculation of the decay $K_1^0 \approx K_{\text{S}}^0 \to \gamma^*\gamma^* \to e^+e^-$ in chiral perturbation theory, which yields the amplitude b_1, G. Ecker and A. Pich obtained [77]

$$\mathcal{P}_{\text{L}}(K_{\text{L}}^0 \to \mu^+\mu^-) \approx 2 \times 10^{-3}. \tag{9.144}$$

Taking into account theoretical uncertainties in their calculation, they conclude that an observation of $\mathcal{P}_{\text{L}} > 5 \times 10^{-3}$ would indicate the existence of a CP-violating mechanism beyond the Standard Model.

Under the assumption of μ-e universality, the decays $K_{\text{L}}^0 \to \mu^+\mu^-$ and $K_{\text{L}}^0 \to e^+e^-$ are induced by the same physical processes. The absorptive (imaginary) part of the amplitude $\mathcal{A}(K_{\text{L}}^0 \to e^+e^-)$ is thus expected to be dominated by the real two-photon intermediate state. From (9.124) it follows that

$$\frac{\mathcal{B}(K_{\text{L}}^0 \to e^+e^-)_{2\gamma}}{\mathcal{B}(K_{\text{L}}^0 \to \mu^+\mu^-)_{2\gamma}} = \left(\frac{m_e}{m_\mu}\right)^2 \frac{\beta_\mu}{\beta_e} \frac{\left[\ln\left(\frac{1+\beta_e}{1-\beta_e}\right)\right]^2}{\left[\ln\left(\frac{1+\beta_\mu}{1-\beta_\mu}\right)\right]^2}, \tag{9.145}$$

i.e.,

$$\mathcal{B}(K_{\text{L}}^0 \to e^+e^-)_{2\gamma} \approx 3.1 \times 10^{-12}. \tag{9.146}$$

The short-distance contribution to the real (dispersive) part of $\mathcal{A}(K_{\text{L}}^0 \to e^+e^-)$ can be expressed as (see (9.119) and (9.120))

$$\frac{\mathcal{B}(K_L^0 \to e^+e^-)_{SD}}{\mathcal{B}(K_L^0 \to \mu^+\mu^-)_{SD}} \approx \frac{\mathcal{B}(K^+ \to e^+\nu_e)}{\mathcal{B}(K^+ \to \mu^+\nu_\mu)} \approx \left(\frac{m_e}{m_\mu}\right)^2. \tag{9.147}$$

We see that $(K_L^0 \to e^+e^-)_{SD}$ is even more suppressed with respect to $(K_L^0 \to mu^+\mu^-)_{SD}$ than $(K_L^0 \to e^+e^-)_{2\gamma}$ is with respect to $(K_L^0 \to \mu^+\mu^-)_{2\gamma}$. The decay $K_L^0 \to e^+e^-$ is, therefore, unlikely to contribute much to our understanding of the short-distance physics.

The first observation of the $K_L^0 \to e^+e^-$ decay mode has recently been made by the E871 collaboration [74]. They observed four $K_L^0 \to e^+e^-$ candidate events, with a predicted background of 0.17 ± 0.10 $K_L^0 \to e^+e^-\gamma$ and $K_L^0 \to e^+e^-e^+e^-$ events. Their observation translates into a branching fraction of $(8.7^{+5.7}_{-4.1}) \times 10^{-12}$.

Supplement: $K^0 \to \pi^-\ell^+\nu_\ell$ and $K^+ \to \pi^0\ell^+\nu_\ell$

The amplitude for the semileptonic decay $K \to \pi\ell\nu_\ell$, where ℓ is μ^+ or e^+, has the form

$$\mathcal{M} = \frac{G_F}{\sqrt{2}} \langle \pi \mid J_h^\beta \mid K \rangle \, \bar{u}(p_\nu)\gamma_\beta(1 - \gamma_5)v(p_\ell) \tag{S9.1}$$

(c.f. (9.85)). Measurements of the π^0-ν_e angular distribution in the decay $K^+ \to \pi^0 e^+\nu_e$ reveal that the scalar and tensor contributions to J_h^β are very small [80] (see Fig. 9.11). Taking also into account that π and K have the same parity, we infer that J_h^β contains only the vector current.

The hadronic amplitude must be formed from the available 4-vectors. It is convenient to write

$$\langle \pi \mid J_h^\beta \mid K \rangle = f_+(q^2)k^\beta + f_-(q^2)q^\beta, \tag{S9.2}$$

where

$$k \equiv k_K + k_\pi, \quad q^2 \equiv (k_K - k_\pi)^2 = (p_\nu + p_\ell)^2. \tag{S9.3}$$

By virtue of the Dirac equation, $\bar{u}(p_\nu)\hat{q}(1 - \gamma_5)v(p_\ell) = m_\ell\bar{u}(p_\nu)(1 + \gamma_5)v(p_\ell)$. The amplitude (S9.1) can thus be written as

$$\mathcal{M} = \frac{G_F}{\sqrt{2}}\{f_+ k^\beta \bar{u}(p_\nu)\gamma_\beta(1-\gamma_5)v(p_\ell) + f_- m_\ell\bar{u}(p_\nu)(1+\gamma_5)v(p_\ell)\}. \tag{S9.4}$$

Summing over the spin states of the leptons, we obtain

$$\begin{aligned}
|\overline{\mathcal{M}}|^2 &= \frac{G_F^2}{2} \text{Tr}\left\{2(1 - \gamma_5)\left[f_+^2\hat{p}_\nu\hat{k}(\hat{p}_\ell + m_\ell)\hat{k} + f_-^2 m_\ell^2\hat{p}_\nu(\hat{p}_\ell + m_\ell)\right.\right. \\
&\quad \left.\left. + f_+ f_-^* m_\ell\hat{p}_\nu\hat{k}(\hat{p}_\ell + m_\ell) + f_- f_+^* m_\ell\hat{p}_\nu(\hat{p}_\ell + m_\ell)\hat{k}\right]\right\} \\
&= 4G_F^2\{f_+^2\left[2(kp_\nu)(kp_\ell) - k^2(p_\nu p_\ell)\right] \\
&\quad + f_-^2 m_\ell^2(p_\nu p_\ell) + 2\text{Re}\,(f_+^* f_-)m_\ell^2(kp_\nu)\}. \tag{S9.5}
\end{aligned}$$

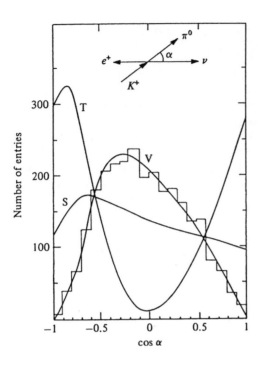

Fig. 9.11. Cosine of the angle between π^0 and ν_e in the dilepton CM system for K_{e3} decay. Predictions for vector (V), scalar (S) and tensor (T) couplings are plotted against data [80]

To derive (S9.5) we used

$$\mathrm{Tr}\left\{\hat{p}_\nu\hat{k}\,\hat{p}_\ell\hat{k}\,\gamma_5\right\} = 4\,\varepsilon_{\alpha\beta\varrho\sigma}\,p_\nu^\alpha k^\beta p_\ell^\varrho k^\sigma = 0, \tag{S9.6}$$

which reflects the fact that there are only three independent 4-momenta ($\varepsilon_{\alpha\beta\varrho\sigma}$ is the completely antisymmetric Levi–Civita tensor: $\varepsilon_{0123} = +1$, $\varepsilon^{0123} = -1$).

In the kaon rest frame, $m_k = E_\pi + E_\ell + E_\nu$ and $\boldsymbol{p}_\ell + \boldsymbol{p}_\nu + \boldsymbol{k}_\pi = 0$. Based on this, the 4-vector products in (S9.5) can be expressed in terms of the center-of-mass (CM) energies of the lepton and pion as follows

$$\begin{aligned}
k^2 &= m_k^2 + m_\pi^2 + 2m_k E_\pi,\\
p_\nu\!\cdot\! p_\ell &= \left\{m_k^2 + m_\pi^2 - m_\ell^2 - 2m_k E_\pi\right\}/2,\\
k\!\cdot\! p_\nu &= \left\{3m_k^2 - m_\pi^2 + m_\ell^2 - 2m_k(2E_\ell + E_\pi)\right\}/2,\\
k\!\cdot\! p_\ell &= \left\{2m_k(2E_\ell + E_\pi) - m_k^2 - m_\pi^2 - m_\ell^2\right\}/2.
\end{aligned} \tag{S9.7}$$

In the CM system, the decay rate is given by

$$\mathrm{d}\Gamma = \frac{|\overline{\mathcal{M}}|^2}{2m_k}\,\mathrm{d}\Phi_3, \tag{S9.8}$$

where $\mathrm{d}\Phi_3$ is the differential element of the three-particle phase space

$$\Phi_3 = \frac{1}{(2\pi)^5}\int \frac{\mathrm{d}^3\boldsymbol{k}_\pi}{2E_\pi}\,\frac{\mathrm{d}^3\boldsymbol{p}_\ell}{2E_\ell}\,\frac{\mathrm{d}^3\boldsymbol{p}_\nu}{2E_\nu}\,\delta^{(3)}(\boldsymbol{k}_\pi + \boldsymbol{p}_\ell + \boldsymbol{p}_\nu)$$

$$\times \delta(m_k - E_\pi - E_\ell - E_\nu)$$

$$= \frac{1}{8(2\pi)^5} \int \frac{\boldsymbol{k}_\pi^2 \, \mathrm{d}|\boldsymbol{k}_\pi| \, \mathrm{d}\Omega_\pi}{E_\pi} \frac{\boldsymbol{p}_\ell^2 \, \mathrm{d}|\boldsymbol{p}_\ell| \, \mathrm{d}\Omega_\ell}{E_\ell}$$

$$\times \frac{\delta(m_k - E_\pi - E_\ell - E_\nu)}{E_\nu}. \tag{S9.9}$$

Since $\boldsymbol{p} \, \mathrm{d}|\boldsymbol{p}| = E \, \mathrm{d}E$ and $\mathrm{d}\Omega = \mathrm{d}(\cos\theta)\mathrm{d}\phi$, this becomes

$$\Phi_3 = \frac{2\pi}{8(2\pi)^5} \int \mathrm{d}\Omega_\pi \mathrm{d}E_\pi \mathrm{d}E_\ell \frac{|\boldsymbol{k}_\pi||\boldsymbol{p}_\ell|}{E_\nu} \, \mathrm{d}(\cos\theta_{\pi e})$$

$$\times \delta(m_k - E_\pi - E_\ell - E_\nu). \tag{S9.10}$$

To simplify (S9.10), note that

$$E_\nu^2 = \boldsymbol{p}_\nu^2 = \boldsymbol{k}_\pi^2 + \boldsymbol{p}_\ell^2 + 2|\boldsymbol{k}_\pi||\boldsymbol{p}_\ell|\cos\theta_{\pi e}, \tag{S9.11}$$

which yields

$$E_\nu \mathrm{d}E_\nu = |\boldsymbol{k}_\pi||\boldsymbol{p}_\ell|\mathrm{d}(\cos\theta_{\pi e}), \tag{S9.12}$$

after differentiation with respect to $\theta_{\pi e}$, while keeping $|\boldsymbol{k}_\pi|$ and $|\boldsymbol{p}_\ell|$ fixed. Substituting (S9.12) in (S9.10) and integrating out the δ-function results in

$$\mathrm{d}\Phi_3 = \frac{1}{32\pi^3} \, \mathrm{d}E_\ell \mathrm{d}E_\pi. \tag{S9.13}$$

When (S9.5) and (S9.7) are substituted in the expression for the decay rate, it reads

$$\frac{\mathrm{d}^2\Gamma}{\mathrm{d}E_\ell \mathrm{d}E_\pi} = \frac{G_F^2 m_k}{8\pi^3} \left\{ f_+^2 \left[\boldsymbol{p}_\pi^2 - (m_k - E_\pi - 2E_\ell)^2 \right. \right.$$

$$+ \frac{m_\ell^2}{4m_k^2}(8m_k E_\ell + 6m_k E_\pi + m_\pi^2 - 3m_k^2) \bigg]$$

$$+ f_-^2 \frac{m_\ell^2}{4m_k^2} \left[m_k^2 + m_\pi^2 - m_\ell^2 - 2m_k E_\pi \right]$$

$$+ 2\mathrm{Re}\,(f_+^* f_-) \frac{m_\ell^2}{4m_k^2} \left[3m_k^2 - m_\pi^2 + m_\ell^2 - 2m_k(2E_\ell + E_\pi) \right] \bigg\}. \tag{S9.14}$$

If the charged lepton is a positron, we can neglect the terms proportional to m_ℓ^2, in which case

$$\mathrm{d}^2\Gamma = \frac{G_F^2 m_k}{8\pi^3} f_+^2 \left\{ \boldsymbol{p}_\pi^2 - (m_k - E_\pi - 2E_e)^2 \right\} \mathrm{d}E_e \mathrm{d}E_\pi. \tag{S9.15}$$

To determine the kinematic limits of phase-space integration, consider

$$\boldsymbol{p}_\ell = -(\boldsymbol{p}_\nu + \boldsymbol{k}_\pi) \longrightarrow \boldsymbol{p}_\ell^2 = \boldsymbol{k}_\pi^2 + \boldsymbol{p}_\nu^2 + 2|\boldsymbol{k}_\pi||\boldsymbol{p}_\nu|\cos\theta_{\pi\nu}. \tag{S9.16}$$

Hence,

$$|\boldsymbol{p}_\ell{}^{\max}_{\min}|^2 = (|\boldsymbol{p}_\nu| \pm |\boldsymbol{k}_\pi|)^2 = (E_\nu \pm |\boldsymbol{k}_\pi|)^2$$
$$= (m_k - E_\pi - E_\ell \pm |\boldsymbol{k}_\pi|)^2, \tag{S9.17}$$

i.e.

$$\underbrace{E_\ell^2}_{\text{max or min}} -m_\ell^2 = (m_k - E_\pi \pm |\boldsymbol{k}_\pi|)^2 + E_\ell^2 - 2(m_k - E_\pi \pm |\boldsymbol{k}_\pi|)E_\ell.$$

Therefore,

$$(E_\ell)^{\max}_{\min} = \frac{(m_k - E_\pi \pm |\boldsymbol{k}_\pi|)^2 + m_\ell^2}{2(m_k - E_\pi \pm |\boldsymbol{k}_\pi|)} \tag{S9.18}$$

and we find that the positron energy at a fixed E_π varies within the range

$$m_k - E_\pi - |\boldsymbol{k}_\pi| \le 2E_e \le m_k - E_\pi + |\boldsymbol{k}_\pi|. \tag{S9.19}$$

Analogously, the pion energy at a given E_ℓ is constrained by

$$(E_\pi)^{\max}_{\min} = \frac{(m_k - E_\ell \pm |\boldsymbol{p}_\ell|)^2 + m_\pi^2}{2(m_k - E_\ell \pm |\boldsymbol{p}_\ell|)}. \tag{S9.20}$$

If the positron mass is neglected,

$$\frac{(m_k - 2E_e)^2 + m_\pi^2}{2(m_k - 2E_e)} \le E_\pi \le \frac{m_k^2 + m_\pi^2}{2m_k}. \tag{S9.21}$$

Expressions (S9.18) and (S9.20) define the contour of a Dalitz plot.

Assuming that $f_+(q^2) \approx$ constant, the distribution (S9.15) can be readily integrated over dE_e from $E_e^{\min}$ to $E_e^{\max}$, with the result

$$d\Gamma = \frac{G_F^2 m_k}{12\pi^3} f_+^2 \left(E_\pi^2 - m_\pi^2\right)^{3/2} dE_\pi. \tag{S9.22}$$

A measurement of the *pion energy spectrum* gives the q^2 dependence of f_+.

Using (S9.21) and rewriting expression (S9.15) in a slightly different form, we have

$$\frac{d\Gamma}{dE_e} = \frac{G_F^2 m_k}{8\pi^3} f_+^2$$
$$\times \int_{E_\pi^{\min}}^{E_\pi^{\max}} \left\{(m_k - 2E_e)\left[2E_\pi - (m_k - 2E_e)\right] - m_\pi^2\right\} dE_\pi. \tag{S9.23}$$

A simple integration yields the *positron energy spectrum*:

$$\frac{d\Gamma}{dE_e} = \frac{G_F^2 m_k}{2\pi^3} f_+^2 \frac{E_e^2(W_e - E_e)^2}{m_k - 2E_e}, \tag{S9.24}$$

where W_e is the maximum energy of the positron in K_{e3} decay (see (S9.29) and (S9.30)).

The *neutrino spectrum* in the rest-frame of the kaon can be constructed from the measured momentum distributions of the pion and electron in the laboratory frame:

$$E_\nu^{\text{CM}} = \frac{m_k^2 - m_\pi^2 - 2(E_e E_\pi - \boldsymbol{p}_e \cdot \boldsymbol{k}_\pi)_{\text{Lab}}}{2m_k}. \tag{S9.25}$$

This expression follows from

$$(k_K - p_\nu)^2_{\text{CM}} = (k_\pi + p_e)^2_{\text{Lab}}. \tag{S9.26}$$

To find the maximum energy of a particle in a three-body decay $M \to m_1 + m_2 + m_3$, let M_{12} be the invariant mass of particles 1 and 2. In the rest system of M,

$$M_{12}^2 = (E_1 + E_2)^2 - (\boldsymbol{p}_1 + \boldsymbol{p}_2)^2 = (M - E_3)^2 - (\boldsymbol{p}_3)^2. \tag{S9.27}$$

Since $\boldsymbol{p}_3 = E_3^2 - m_3^2$, we have

$$M_{12}^2 = M^2 - 2ME_3 + m_3^2 \longrightarrow E_3 = \frac{M^2 + m_3^2 - M_{12}^2}{2M}. \tag{S9.28}$$

E_3 has the maximum value for $M_{12}^{\min} = m_1 + m_2$. Therefore,

$$W_3 \equiv E_3^{\max} = \frac{M^2 + m_3^2 - (m_1 + m_2)^2}{2M}. \tag{S9.29}$$

In our case one of the particles is a massless neutrino. We thus obtain the following kinematic limits

$$m_\pi \leq E_\pi \leq \frac{m_k^2 + m_\pi^2 - m_\ell^2}{2m_k}, \quad m_\ell \leq E_\ell \leq \frac{m_k^2 + m_\ell^2 - m_\pi^2}{2m_k}. \tag{S9.30}$$

The K_{e3} decay rate is therefore given by

$$\Gamma = \frac{G_F^2 m_k}{12\pi^3} f_+^2 \int_{m_\pi}^{m_k(1+a)/2} E_\pi^3 \left(1 - \frac{m_\pi^2}{E_\pi^2}\right)^{3/2} dE_\pi, \tag{S9.31}$$

where $a \equiv m_\pi^2/m_k^2$. Setting $1 - m_\pi^2/E_\pi^2 = x^2$ and integrating by parts, we obtain

$$\Gamma = \frac{G_F^2 m_k^5}{768\pi^3} f_+^2 \left\{ (1-a)^3(1+a) - 6a(1-a)(1+a) - 12a^2 \ln a \right\}. \tag{S9.32}$$

If only the leading orders in a are kept, this simplifies to

$$\Gamma = \frac{G_F^2 m_k^5}{768\pi^3} f_+^2 \left(1 - \frac{8m_\pi^2}{m_k^2}\right). \tag{S9.33}$$

Of special interest is the *muon polarization* in the decay $K^+ \to \pi^0 \ell^+ \nu_\ell$. Summing over the spin states of the neutrino (see the text preceding expressions (9.133) and (9.134) in Sect. 9.3), we obtain

$$\begin{aligned}
\sum_{s_\nu} |\mathcal{M}|^2 &= \frac{G_F^2}{4} \text{Tr}\big[(f_+ k^\beta + f_- q^\beta)(f_+^* k^\lambda + f_-^* q^\lambda) \\
&\quad \times \hat{p}_\nu \gamma_\beta (1 - \gamma_5)(\hat{p}_\mu - m_\mu)(1 + \gamma_5 \hat{s}_\mu)\gamma_\lambda(1 - \gamma_5)\big] \\
&= \frac{G_F^2}{2} \text{Tr}\big[(f_+ k^\beta + f_- q^\beta)(f_+^* k^\lambda + f_-^* q^\lambda)(1 - \gamma_5) \\
&\quad \times \hat{p}_\nu \gamma_\beta(\hat{p}_\mu + m_\mu \hat{s}_\mu)\gamma_\lambda\big].
\end{aligned} \tag{S9.34}$$

To derive (S9.34) we used $p_\mu \cdot s_\mu = 0$ (see (9.134)). A straightforward evaluation of the traces yields

$$\sum_{s_\nu} |\mathcal{M}|^2 = \frac{1}{2}|\overline{\mathcal{M}}|^2 + 2G_F^2 m_\mu \left\{ f_+^2 \left[2(k \cdot p_\nu)(k \cdot s_\mu) - k^2(p_\nu \cdot s_\mu) \right] \right.$$
$$+ f_-^2 \left[2(q \cdot p_\nu)(q \cdot s_\mu) - q^2(p_\nu \cdot s_\mu) \right]$$
$$+ 2\mathrm{Re}\,(f_+ f_-^*) \left[(k \cdot p_\nu)(q \cdot s_\mu) + (q \cdot p_\nu)(k \cdot s_\mu) - (k \cdot q)(p_\nu \cdot s_\mu) \right]$$
$$\left. + 2\mathrm{Im}\,(f_+ f_-^*)\, \varepsilon_{\beta\lambda\alpha\tau}\, k^\beta q^\lambda p_\nu^\alpha s_\mu^\tau \right\} \tag{S9.35}$$

with $|\overline{\mathcal{M}}|^2$ given by (S9.5). In the kaon rest frame,

$$q \cdot s_\mu = p_\nu \cdot s_\mu = E_\nu \frac{\boldsymbol{p}_\mu \cdot \boldsymbol{s}_\mu}{m_\mu} - \boldsymbol{p}_\nu \cdot \boldsymbol{s}_\mu - \frac{(\boldsymbol{p}_\mu \cdot \boldsymbol{s}_\mu)(\boldsymbol{p}_\mu \cdot \boldsymbol{p}_\nu)}{m_\mu(E_\mu + m_\mu)},$$

$$k \cdot s_\mu = (2k_K - p_\mu - p_\nu) \cdot s_\mu = 2m_k \frac{\boldsymbol{p}_\mu \cdot \boldsymbol{s}_\mu}{m_\mu} - \boldsymbol{p}_\nu \cdot \boldsymbol{s}_\mu \tag{S9.36}$$

($\boldsymbol{s}_\mu$ is the spin unit vector of the muon in its rest frame) and

$$\varepsilon_{\beta\lambda\alpha\tau}\, k^\beta q^\lambda p_\nu^\alpha s_\mu^\tau = \varepsilon_{\beta\lambda\alpha\tau}(2k_K - p_\mu - p_\nu)^\beta (p_\mu + p_\nu)^\lambda p_\nu^\alpha s_\mu^\tau$$
$$= 2\varepsilon_{\beta\lambda\alpha\tau}\, k^\beta p_\mu^\lambda p_\nu^\alpha s_\mu^\tau$$
$$= 2m_k \varepsilon_{i0jk}\, p_\mu^i p_\nu^j s_\mu^k = 2m_k(\boldsymbol{p}_\mu \times \boldsymbol{p}_\nu) \cdot \boldsymbol{s}_\mu. \tag{S9.37}$$

Each term in (S9.35), with the exception of the last one, contains components of the muon polarization in the decay plane. The last term gives rise to a polarization component normal to that plane. Since this term changes sign under time reversal, its presence would violate T invariance. Indeed, the time-reversed amplitude of (S9.1) and (S9.2)

$$\mathcal{M} \xrightarrow{T} \frac{G_F}{\sqrt{2}} \left[f_+^*(q^2)k_\beta + f_-^*(q^2)q_\beta \right] \bar{u}(p_\nu)\gamma^\beta(1 - \gamma_5)v(p_\ell) \tag{S9.38}$$

also leads to (S9.35), except that in this case the last term is $\propto \mathrm{Im}\,(f_+^* f_-) = -\mathrm{Im}\,(f_+ f_-^*)$. If the decay is caused by an interaction that is T invariant, $\mathrm{Im}\,(f_+ f_-^*) = 0$ (f_+ and f_- have the same phase). To obtain (S9.38) we used Table E.2 in Appendix E and the fact that the operation of time reversal implies charge conjugation.

We now define

$$\alpha = \frac{m_\mu}{2}\left[1 - \frac{f_-}{f_+} \right] = \frac{m_\mu}{2}\left[1 - \xi(q^2) \right], \quad \Re \equiv 2\,\mathrm{Re}\,\alpha, \quad \Im \equiv 2\,\mathrm{Im}\,\xi, \tag{S9.39}$$

and rewrite (S9.35) as

$$\sum_{s_\nu} |\mathcal{M}|^2 = 8G_F^2 f_+^2 \left[A + B \cdot s_\mu \right], \tag{S9.40}$$

where

$$A = \left[2(k_K \cdot p_\mu) - \Re m_\mu \right](k_K \cdot p_\nu) - (m_k^2 - \alpha^2)(p_\mu \cdot p_\nu), \tag{S9.41}$$

$$B = \left[2m_\mu(k_K \cdot p_\nu) - \Re(p_\mu \cdot p_\nu)\right]k_K^\sigma - \left[m_\mu(m_k^2 + \alpha^2) - \Re(k_K \cdot p_\mu)\right]p_\nu^\sigma$$
$$+ \frac{\Im}{4}\varepsilon_{\beta\lambda\alpha\tau}k^\beta q^\lambda p_\nu^\alpha s_\mu^\tau. \tag{S9.42}$$

It can be readily verified, using (S9.36) and (S9.37), that in the kaon rest frame

$$B \cdot s_\mu = \left\{a(\xi)\boldsymbol{k}_\pi + \left[b(\xi) + a(\xi)\left(\frac{\boldsymbol{k}_\pi \cdot \boldsymbol{p}_\mu}{E_\mu + m_\mu} + m_k - E_\pi\right)\right]\boldsymbol{p}_\mu \right.$$
$$\left. + \frac{\Im}{2}\,m_k(\boldsymbol{k}_\pi \times \boldsymbol{p}_\mu)\right\} \cdot s_\mu. \tag{S9.43}$$

In the above expression,

$$a(\xi) = -m_\mu\left[m_k^2 + \alpha^2 - \frac{\Re}{m_\mu}m_k E_\mu\right],$$
$$b(\xi) = m_k^2\left[2E_\nu - \frac{\Re}{m_\mu}(W_\pi - E_\pi)\right]. \tag{S9.44}$$

(W_π is the maximum energy of the pion; see (S9.29) and (S9.30)).

From (S9.43) it follows immediately that the muon polarization vector is given by [81]

$$\boldsymbol{P}(\xi) = \frac{\boldsymbol{A}(\xi)}{|\boldsymbol{A}(\xi)|},$$

$$\boldsymbol{A} = a(\xi)\boldsymbol{k}_\pi + \left[b(\xi) + a(\xi)\left(\frac{\boldsymbol{k}_\pi \cdot \boldsymbol{p}_\mu}{E_\mu + m_\mu} + m_k - E_\pi\right)\right]\boldsymbol{p}_\mu \tag{S9.45}$$
$$+ \frac{\Im}{2}\,m_k(\boldsymbol{k}_\pi \times \boldsymbol{p}_\mu).$$

Since the pion and K meson are spinless and the neutrino has a definite helicity, the muon is completely polarized. The direction of its polarization is fixed by the kinematics of the decay. The nonvanishing mass of the muon gives rise to a polarization component along the pion momentum. As we explained earlier, if the decay is T invariant, ξ is real and the polarization vector lies entirely in the decay plane.

9.4 Direct CP Violation (ε')

The parameter ε', defined by (3.75), determines the amount of direct CP violation in $\Delta S = 1$ transitions:

$$\varepsilon' = \frac{1}{\sqrt{2}}\frac{\operatorname{Re}A_2}{\operatorname{Re}A_0}\left[\frac{\operatorname{Im}A_2}{\operatorname{Re}A_2} - \frac{\operatorname{Im}A_0}{\operatorname{Re}A_0}\right]e^{i(\pi/2 + \delta_2 - \delta_0)}. \tag{9.148}$$

Any CP-violating observable must involve an interference of two amplitudes. In the case of ε', the interfering amplitudes are

$$\langle \pi\pi, \ I = \alpha | \hat{H}_w | K^0 \rangle \equiv A_\alpha e^{i\delta_\alpha}, \quad \alpha = 0, 2, \tag{9.149}$$

where $|\pi\pi, I = 0, 2\rangle$ are the $I = 0$ and $I = 2$ isospin states of the 2π system.

Remarkably, the phase factor in (9.148) is very close to $\pi/4$, i.e., the phase of ε' is almost the same as that of ε. Using (3.74), (3.81) and (3.82), we can express the parameter ε, which measures the interference between $K^0 \to \pi\pi$ and $K^0 \to \bar{K}^0 \to \pi\pi$, as follows:

$$\varepsilon \approx \frac{1}{\sqrt{2}} e^{i\phi_{\pi\pi}} \left[\frac{\mathrm{Im}\, M_{12}}{2\mathrm{Re}\, M_{12}} + \frac{\mathrm{Im}\, A_0}{\mathrm{Re}\, A_0} \right]. \tag{9.150}$$

The expressions for ε' and ε are independent of phase convention. To show this we redefine the phases of the strangeness eigenstates K^0 and $\bar{K}^0$: $|K^0\rangle \to e^{-i\lambda}|K^0\rangle$, $|\bar{K}^0\rangle \to e^{i\lambda}|\bar{K}^0\rangle$. For $|\lambda| \ll 1$,

$$\mathrm{Im}\, \langle K^0 | \hat{H}_w | \bar{K}^0 \rangle \to \mathrm{Im}\, \left[e^{2i\lambda} \langle K^0 | \hat{H}_w | \bar{K}^0 \rangle \right]$$

$$\approx \mathrm{Im}\, \langle K^0 | \hat{H}_w | \bar{K}^0 \rangle + 2\lambda \mathrm{Re}\, \langle K^0 | \hat{H}_w | \bar{K}^0 \rangle, \tag{9.151}$$

i.e.,

$$\frac{\mathrm{Im}\, M_{12}}{\mathrm{Re}\, M_{12}} \to \frac{\mathrm{Im}\, M_{12}}{\mathrm{Re}\, M_{12}} + 2\lambda. \tag{9.152}$$

Similarly,

$$\frac{\mathrm{Im}\, A_\alpha}{\mathrm{Re}\, A_\alpha} \to \frac{\mathrm{Im}\, A_\alpha}{\mathrm{Re}\, A_\alpha} - \lambda, \tag{9.153}$$

which proves the above assertion. In the absence of $\Delta I = 3/2$ transitions, $\varepsilon' \propto \mathrm{Im}\, A_0/\mathrm{Re}\, A_0$. Since this expression is not phase-convention invariant, ε' must vanish for $A_2 = 0$. Therefore, to determine ε', one has to calculate the phases of both A_0 and A_2.

Using the experimental ratio of $\Delta I = 3/2$ and $\Delta I = 1/2$ amplitudes

$$\frac{\mathrm{Re}\, A_2}{\mathrm{Re}\, A_0} \approx \sqrt{\frac{\Gamma(K^+ \to \pi^+\pi^0)}{\Gamma(K_S \to \pi\pi)}} \equiv \omega \approx \frac{1}{22} \tag{9.154}$$

and the measured values $(\pi/2 + \delta_2 - \delta_0) \approx \pi/4$, $\phi_{\pi\pi} \approx \pi/4$, we obtain

$$\frac{\varepsilon'}{\varepsilon} \approx \frac{\omega}{\sqrt{2}|\varepsilon|} \left[\frac{\mathrm{Im}\, A_2}{\mathrm{Re}\, A_2} - \frac{\mathrm{Im}\, A_0}{\mathrm{Re}\, A_0} \right] \equiv \frac{-\omega}{\sqrt{2}|\varepsilon|} \frac{\mathrm{Im}\, A_0}{\mathrm{Re}\, A_0}(1 - \Omega), \tag{9.155}$$

where $\Omega \equiv \omega^{-1} \mathrm{Im}\, A_2/\mathrm{Im}\, A_0$. For reasons unrelated to CP violation, the phase $\mathrm{Im}\, A_2/\mathrm{Re}\, A_2$ is enhanced due to the smallness of $\mathrm{Re}\, A_2$. This means that any interaction that gives rise to A_2 is 22 times more effective in producing ε' than the one which contributes only to A_0. From

$$\Gamma_S = \left| \langle I = 0 | \hat{H}_w | K_S^0 \rangle + \langle I = 2 | \hat{H}_w | K_S^0 \rangle \right|^2 \frac{\sqrt{1 - 4m_\pi^2/m_k^2}}{16\pi m_k} \tag{9.156}$$

it follows (see (9.149)) that

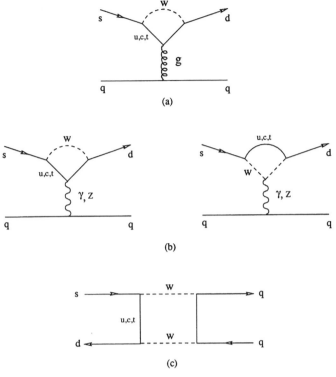

Fig. 9.12. Gluonic **(a)** and electroweak **(b,c)** "penguins" giving rise to ε'

$$\text{Re } A_0 \approx 3.34 \times 10^{-7} \text{ GeV.} \qquad (9.157)$$

As first noted by F. Gilman and M. Wise [78], a nonzero value of ε' could arise from the diagram in Fig. 9.12a, whimsically named "gluonic penguin". The diagram, which is the main source of CP violation in $\Delta S = 1$ transitions, describes the $\Delta I = 1/2$ decay $s \to d + g$, where g is a gluon. Since this graph turns an s quark into a d quark, it can only change isospin by a unit of $1/2$, and thus cannot lead to an $I = 2$ final state.

When g is replaced by a photon (or a Z^0), the resulting "electroweak penguin" diagram (Fig. 9.12b) generates a phase in A_2. To see this, note that the electromagnetic current

$$\bar{q}\gamma^\alpha Q_q q = \bar{q}\gamma^\alpha \frac{1 + \gamma_5}{2} Q_q q + \bar{q}\gamma^\alpha \frac{1 - \gamma_5}{2} Q_q q. \qquad (9.158)$$

(Q_q is the charge of quark q) is a mixture of $I = 1$ and $I = 0$ components:

$$\bar{q} Q_q q = \frac{2\bar{u}u - \bar{d}d - \bar{s}s}{3} = \frac{\bar{u}u - \bar{d}d}{2} + \frac{\bar{u}u + \bar{d}d - 2\bar{s}s}{6}. \qquad (9.159)$$

The $I = 1$ component of the electromagnetic current and the $s \to d$ transition current, which changes isospin by a unit of $1/2$, can induce a $\Delta I = 3/2$ transition and thus create an $I = 2$ final state (recall that $I_k = 1/2$).

At lowest order, the diagrams in Fig. 9.12a,b yield the following effective hamiltonians [78, 79]:

$$\hat{H}_w(g) \approx -\frac{G_F}{\sqrt{2}} \frac{\alpha_s}{12\pi} \left\{ V_{ud}^* V_{us} \ln \frac{m_c^2}{\mu^2} + V_{td}^* V_{ts} \ln \frac{m_t^2}{m_c^2} \right\} \bar{d}\gamma_\mu (1 - \gamma_5) \lambda^a s$$
$$\times \sum_{q=u,d,s} \bar{q}\gamma^\mu \lambda_a q \qquad (9.160)$$

and

$$\hat{H}_w(\gamma) \approx -\frac{G_F}{\sqrt{2}} \frac{2\alpha}{9\pi} \left\{ V_{ud}^* V_{us} \ln \frac{m_c^2}{\mu^2} + V_{td}^* V_{ts} \ln \frac{m_t^2}{m_c^2} \right\} \bar{d}\gamma_\mu (1 - \gamma_5) s$$
$$\times \sum_{q=u,d,s} \bar{q}\gamma^\mu Q_q q. \qquad (9.161)$$

Here we show only the logarithmic $m_{c,t}$ dependences. In (9.160) and (9.161), μ is a typical hadronic mass scale, λ^a are color SU(3) matrices and α_s is the strong coupling constant ($\alpha_s \approx 10^{-1}$).

Since $V_{ud}^* V_{us}$ is purely real, the contribution to ε' comes only from $\mathrm{Im}\,(V_{td}^* V_{ts}) = -A^2 \lambda^5 \eta$:

$$\mathrm{Im}\,A_0 \approx \frac{G_F}{\sqrt{2}} A^2 \lambda^5 \eta \, \langle \pi\pi, I = 0 \,|\, C_p \hat{O}_p \,|\, K^0 \rangle \qquad (9.162)$$

with

$$C_p \equiv \frac{\alpha_s}{12\pi} \ln \frac{m_t^2}{m_c^2}, \quad \hat{O}_p \equiv \bar{d}\gamma_\mu (1 - \gamma_5) \lambda^a s \sum_{q=u,d,s} \bar{q}\gamma^\mu \lambda_a q \qquad (9.163)$$

and similarly for A_2. The photonic penguin is suppressed with respect to its gluonic counterpart by the factor of α/α_s. However, the phase $\mathrm{Im}\,A_2/\mathrm{Re}\,A_2$ is enhanced due to the smallness of $\mathrm{Re}\,A_2$, as mentioned above.

A full gauge-invariant calculation includes Z^0-penguins and box diagrams (Fig. 9.12c). The detailed calculations are very complicated. In essence, one evaluates the relevant Feynman diagrams to obtain an effective hamiltonian for $K^0 \to \pi\pi$ (as was discussed for K^0-$\bar{K}^0$ mixing), extracts its operator structure and then tries to estimate the hadronic matrix elements of these operators. The amplitudes $\mathrm{Im}\,A_{0,2}$ are calculated from the effective hamiltonian for $\Delta S = 1$ transitions

$$\hat{H}_{\mathrm{eff}}^{\Delta S=1} = -\frac{G_F}{\sqrt{2}} \mathrm{Im}\,(V_{td}^* V_{ts}) \sum_{i=1}^{8} C_i(\mu) \hat{O}_i \qquad (9.164)$$

derived using *renormalization group* (RG) equations. The coefficient functions $C_i(\mu)$, which are governed by the short-distance structure of QCD ($\mu \leq E \leq m_w$), have been calculated at next-to-leading order in perturbation theory.

The main difficulty lies in the evaluation of the hadronic matrix elements

$$\langle O_i \rangle_{0,2} \equiv \langle \pi\pi, I = 0, 2 \mid \hat{O}_i \mid K^0 \rangle, \tag{9.165}$$

which describe the long-distance physics ($0 \le E \le \mu$), and therefore cannot be treated perturbatively. The dominant contribution comes from two operators, $\hat{O}_6$ (gluonic penguin) and $\hat{O}_8$ (electroweak penguin), which are responsible for the $\Delta I = 1/2$ and $\Delta I = 3/2$ transitions, respectively. Expression (9.155) thus simplifies to

$$\frac{\varepsilon'}{\varepsilon} = \frac{\omega G_F}{2|\varepsilon| \mathrm{Re}\, A_0} \, \mathrm{Im}\, (V_{td}^* V_{ts}) C_6 \langle O_6 \rangle_0 \left[1 - \omega^{-1} \frac{C_8 \langle O_8 \rangle_2}{C_6 \langle O_6 \rangle_0} \right]. \tag{9.166}$$

We should also mention the small breaking of isospin invariance that gives rise to π-η and π-η' mixing. As a consequence, $K^0 \to \pi^0 \pi^0$ will receive contributions from diagrams in Fig. 9.13. The amplitudes $A(K^0 \to \pi^0 \eta)$ and $A(K^0 \to \pi^0 \eta')$ get imaginary parts from the gluonic penguin in Fig. 9.12a. This induces, via the processes in Fig. 9.13, an imaginary part in $A(K^0 \to \pi^0\pi^0)$ and thus in both the $I = 0$ and $I = 2$ components of the amplitude $A(K^0 \to 2\pi)$ (see (3.55) and (3.76)).

Fig. 9.13. Isospin-breaking contribution to $K^0 \to \pi^0 \pi^0$

The electroweak-penguin (ewp) and isospin-breaking (IB) effects modify the purely gluonic contribution by (see (9.155))

$$\Omega = \Omega_{ewp} + \Omega_{\eta+\eta'}, \quad \Omega_{\eta+\eta'} = \omega^{-1} \frac{(\mathrm{Im}\, A_2)_{IB}}{\mathrm{Im}\, A_0}. \tag{9.167}$$

The Standard Model prediction for ε'/ε suffers from large hadronic uncertainties, aggravated by substantial cancellations between the $I = 0$ and $I = 2$ contributions (the $\hat{O}_6$ contribution to ε'/ε is positive and only weakly m_t-dependent, whereas that of $\hat{O}_8$ is negative and shows a strong m_t dependence). This renders a precise calculation of this quantity impossible at present. A recent comprehensive review of the subject that provides access to the theoretical literature can be found in [54]. Here we quote their own result,

$$-2.1 \times 10^{-4} \le \varepsilon'/\varepsilon \le 13.2 \times 10^{-4}, \tag{9.168}$$

but stress that some other authors obtain a much wider range of values for ε'/ε.

Appendices

A *CP* Properties of $K \to 2\pi$ and $K \to 3\pi$

The pion has spin zero and negative intrinsic parity (it is a *pseudoscalar*): $J_\pi^P = 0^-$. Conservation of angular momentum therefore requires the spin of the kaon to be equal to the relative orbital angular momentum, ℓ, of the two pions in the decay $K \to 2\pi$: $J_k = J_{2\pi} = \ell$. The parity of the 2π state is

$$P_{2\pi} = (P_\pi)^2(-1)^\ell = (-1)^2(-1)^\ell = (-1)^\ell. \tag{A.1}$$

To find the allowed values of ℓ, note that the identity of the final state pions in the $K^0 \to 2\pi^0$ decay implies that their wavefunction is symmetric, which means that ℓ must be even: $\ell = 0, 2, 4, \ldots$. Hence,

$$J_{2\pi}^P = 0^+, \; 2^+, \; 4^+, \; \ldots . \tag{A.2}$$

The 3π system in the charged kaon decay can be regarded as consisting of two parts: a 2π state containing the pions of like charge ($\pi^+\pi^+$ or $\pi^-\pi^-$), plus the remaining pion. If we denote the relative orbital angular momentum in the 2π state by ℓ, and that of the third pion relative to 2π state by L, then the spin of the 3π system is constrained by

$$|\ell - L| \leq J_{3\pi} \leq |\ell + L|. \tag{A.3}$$

The parity of the final state is

$$P_{3\pi} = (P_\pi)^3(-1)^\ell(-1)^L = (-1)^{L+1}, \tag{A.4}$$

because ℓ must be even for two indentical bosons. Since the sum of the masses of the three pions is close to the kaon mass, the decays into 3π states with $\ell > 0$ are supressed by kinematics. We can thus write

$$J_{3\pi}^P = L^{(-1)^{L+1}} = 0^-, \; 1^+, \; 2^-, \; \ldots . \tag{A.5}$$

From (A.2) and (A.5) we infer that

$$J_{3\pi}^P = 0^-, \; 2^-, \; 4^-, \; \ldots . \tag{A.6}$$

Measurements of the charged 3π decay modes favor $J_k = 0$. This result is corroborated by the absence of the decay $K^\pm \to \pi^\pm + \gamma$, which is forbidden if the $K^\pm$ spin is zero. Furthermore, the muon polarization in the decay

$K^{\pm} \to \mu^{\pm} + \nu_{\mu}$ is the same as in $\pi^{\pm} \to \mu^{\pm} + \nu_{\mu}$, thus indicating that $J_k = J_{\pi} = 0$. Therefore,

$$J_{2\pi}^P = 0^+, \ J_{3\pi}^P = 0^-. \tag{A.7}$$

The kaon parity can be determined from the reaction $K^- + {}^4\text{He} \to {}^4\text{H}_A + \pi^0$ (K^- capture from an atomic S state), where the hypernucleus ${}^4\text{H}_A$ contains a bound Λ hyperon in place of a neutron. Measurements of ${}^4\text{H}_A$ weak decay modes suggest that $J = \ell = 0$ on both sides of the reaction. Hence $J_k^P = 0^-$, if the parity of the Λ-particle is defined to be positive, like that of the nucleon.

Application of the charge conjugation operator $\hat{C}$ interchanges the π^+ and π^- in the decay $K^0 \to \pi^+\pi^-$. In this case the operation is equivalent to space inversion,

$$\hat{C}\Psi_{\pi^+\pi^-} = \hat{P}\Psi_{\pi^+\pi^-} = (-1)^{\ell}\Psi_{\pi^+\pi^-}, \tag{A.8}$$

and thus

$$\hat{C}\hat{P}\Psi_{\pi^+\pi^-} = (-1)^{2\ell}\Psi_{\pi^+\pi^-} = +\Psi_{\pi^+\pi^-}. \tag{A.9}$$

For the $\pi^0\pi^0$ final state

$$\hat{C}\hat{P}\Psi_{\pi^0\pi^0} = \hat{P}\Psi_{\pi^0\pi^0} = (-1)^{\ell}\Psi_{\pi^0\pi^0} = +\Psi_{\pi^0\pi^0} \tag{A.10}$$

because two identical bosons must be in an overall symmetric state. If we also take into account that the kaon is spinless, conservation of angular momentum requires $\ell = 0$ in the 2π state. Hence,

$$\hat{C}\Psi_{2\pi} = \hat{P}\Psi_{2\pi} = +\Psi_{2\pi}, \tag{A.11}$$

Regarding the decay $K^0 \to 3\pi^0$, we have shown that for a system of three pions, at least two of which are identical, $P = -1$:

$$\hat{C}\hat{P}\Psi_{3\pi^0} = -\Psi_{3\pi^0}. \tag{A.12}$$

As for the charged 3π decay mode, for $K^0 \to \pi^+\pi^-\pi^0$ we take the relative orbital angular momentum in the $\pi^+\pi^-$ state to be ℓ, and of the π^0 relative to the $\pi^+\pi^-$ centre of mass to be L. Now,

$$\hat{C}\hat{P}\Psi_{\pi^+\pi^-\pi^0} = \hat{C}\hat{P}\{\Psi_{\pi^+\pi^-}(\ell)\Psi_{\pi^0}(L)\} = +\Psi_{\pi^+\pi^-}\hat{P}\Psi_{\pi^0}(L)$$
$$= (-1)^{L+1}\Psi_{\pi^+\pi^-\pi^0}. \tag{A.13}$$

We see that, in contrast to the $\pi^+\pi^-$, $\pi^0\pi^0$ and $3\pi^0$ final states, the $\pi^+\pi^-\pi^0$ system does not have a well-defined CP eigenvalue: $CP = (-1)^{L+1}$. The kaon is spinless, and so the total angular momentum in the 3π-system must be zero, which means that L must be matched by $\ell = L$. If $L = 0$ or an even integer, then $CP = -1$; if L is odd, $CP = +1$. As explained above, states with $\ell > 0$ are supressed by kinematics. Therefore, the decay $K^0 \to \pi^+\pi^-\pi^0$ is expected to be dominated by the CP-allowed decay $K_2^0 \to \pi^+\pi^-\pi^0$ with $L = \ell = 0$ and $CP = -1$. The K_1^0 may decay into the kinematics-supressed and CP-allowed final state with $L = \ell = 1$ and $CP = +1$, or into the kinematics-favored and CP-forbidden state with $L = \ell = 0$ and $CP = -1$.

The charge conjugation operator $\hat{C}$ changes the sign of all charges, including the charge of the sources and thus of the electromagnetic fields they produce:

$$A_\mu(x) \longrightarrow -A_\mu(x), \quad \text{under } \hat{C}. \tag{A.14}$$

If we consider an electromagnetic field to be a collection of photons, then the electromagnetic potential $A_\mu(x)$ can be taken to represent the photon wavefunction. From (A.14) we infer that the photon has odd charge-conjugation parity

$$\hat{C}|\gamma\rangle = -|\gamma\rangle, \tag{A.15}$$

i.e., $|\gamma\rangle$ is an eigenstate of $\hat{C}$ with eigenvalue $C = -1$. Since C is multiplicative,

$$\hat{C}|n\gamma\rangle = \hat{C}|\gamma_1\rangle|\gamma_2\rangle \dots |\gamma_n\rangle = (-1)^n|n\gamma\rangle. \tag{A.16}$$

The π^0 decays predominantly to two photons. Hence,

$$\hat{C}|\pi^0\rangle = +|\pi^0\rangle, \quad \hat{C}\hat{P}|\pi^0\rangle = -|\pi^0\rangle. \tag{A.17}$$

B Forward Scattering Amplitude and the Optical Theorem

Suppose an incident beam of particles, represented by a plane wave traveling in the z direction, $\Psi_i = \mathrm{e}^{\mathrm{i}kz}$, impinges normally on a thin slab of material (see Fig. B.1). The scattered beam at large r is a spherical wave propagating outward from each of the scatterers:

$$\Psi_{\mathrm{sc}} = \mathfrak{f}(\theta)\frac{\mathrm{e}^{\mathrm{i}kr}}{r}, \tag{B.1}$$

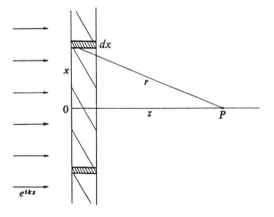

Fig. B.1. A plane wave impinges on a thin slab of material

where $f(\theta)$ is the scattering amplitude. Note that in (B.1) there is no dependence on azimuthal angle ϕ because the z component of the angular momentum is zero. We have also ignored the factor e^{-iEt}, since for coherent scattering the slab as a whole takes up the recoil, and therefore very little energy is exchanged in the process:

$$E_1 - E_2 = \frac{P_{\text{slab}}^2}{2M_{\text{slab}}} \ll P_{\text{slab}} \equiv P_1 - P_2.$$

The scattered wave must have a $1/r$ dependence to ensure that the rate of scattered particles that pass through a spherical surface centred around $r = 0$ is independent of r:

$$\mathcal{R} \equiv |\Psi_{\text{sc}}|^2 v_{\text{rel}}\, r^2 d\Omega = v_{rel}|f(\theta)|^2 d\Omega \tag{B.2}$$

$\mathcal{R}$ is the number of scattered particles per unit time through the surface $r^2 d\Omega$ and v_{rel} the relative velocity of outgoing particles and the target. According to the definition of a cross-section,

$$\mathcal{R} \equiv \text{flux} \cdot d\sigma = v_{\text{rel}}\, d\sigma, \tag{B.3}$$

for a scattering centre per unit area and the incident flux of particles

$$\text{flux} = v_i \Psi_i \Psi_i^* = v_i \approx v_{\text{rel}}. \tag{B.4}$$

Therefore, the differential cross-section for elastic scattering reads

$$\frac{d\sigma}{d\Omega} = |f(\theta)|^2. \tag{B.5}$$

For a slab of material of thickness l, containing N scatterers per unit volume, the amplitude $\mathcal{A}_{\text{sc}}(z)$ of the scattered wave at a distance z is given by

$$\mathcal{A}_{\text{sc}}(z) = \int_0^\infty [2\pi x dx Nl]\, f(\omega, \theta)\, \frac{e^{ikr}}{r}. \tag{B.6}$$

The expression in the square bracket represents the number of scatterers in a ring at radius x. Since $r = \sqrt{x^2 + z^2}$, we can write

$$\mathcal{A}_{sc}(z) = 2\pi Nl \int_z^\infty dr\, f(\omega, \theta) e^{ikr}. \tag{B.7}$$

Defining $u = f(\omega, \theta)$, $dv = e^{ikr} dr$ and integrating by parts, results in

$$\mathcal{A}_{\text{sc}}(z) = \frac{2\pi Nl}{ik} \left\{ f(\omega, \theta)\, e^{ikr}\Big|_z^\infty - \int_z^\infty e^{ikr} df(\omega, \theta) \right\}. \tag{B.8}$$

But $e^{ikr} \overset{r\to\infty}{\longrightarrow} 0$, and so the first term in (B.8) gives

$$\frac{i2\pi Nl}{k} f(\omega, \theta) e^{ikz}.$$

For the scattering in the forward direction ($\theta \to 0$) the second term does not contribute. The scattering amplitude at $\theta = 0$ is, therefore, determined by the first term:

$$A_{\text{sc}}(z) = \frac{\mathrm{i}2\pi N l}{k}\, \mathfrak{f}(\omega, 0) \mathrm{e}^{\mathrm{i}kz}. \tag{B.9}$$

The total amplitude at large z is then

$$A(z) = A_{\text{in}}(z) + A_{\text{sc}}(z) = \mathrm{e}^{\mathrm{i}kz} + \frac{\mathrm{i}2\pi N l}{k}\, \mathfrak{f}(\omega, 0) \mathrm{e}^{\mathrm{i}kz}, \tag{B.10}$$

i.e.

$$A(z) = \left[1 + \frac{\mathrm{i}2\pi N l}{k}\, \mathfrak{f}(\omega, 0)\right] \mathrm{e}^{\mathrm{i}kz}. \tag{B.11}$$

Let us assume that the slab is a homogeneous material of refractive index $n(\omega)$. The wave number in the material is nk, whereas in vacuum it is k. The above amplitude can thus be expressed as

$$A(z) = \mathrm{e}^{\mathrm{i}k(z-l)+\mathrm{i}nkl} = \mathrm{e}^{\mathrm{i}kz}\mathrm{e}^{\mathrm{i}k(n-1)l}, \tag{B.12}$$

which shows that the particle picks up an extra phase $\phi_{\text{ext}} \equiv k(n-1)l$. For $\phi_{\text{ext}} \ll 1$, a Taylor expansion of the second exponential in (B.12) gives

$$A(z) \approx \mathrm{e}^{\mathrm{i}kz}\left[1 + \mathrm{i}k(n-1)l\right]. \tag{B.13}$$

Comparing (B.11) and (B.13) we infer that

$$n(\omega) = 1 + \frac{2\pi N}{k^2}\, \mathfrak{f}(\omega, 0) = 1 + \frac{2\pi N c^2}{\omega^2}\, \mathfrak{f}(\omega, 0), \tag{B.14}$$

where $\omega = kc$. Expression (B.14) relates the forward scattering amplitude $\mathfrak{f}(\omega, 0)$ with the index of refraction $n(\omega)$.

We next derive the *optical theorem*

$$\sigma_{\text{tot}} = \frac{4\pi}{k}\mathrm{Im}\, \mathfrak{f}(\omega, 0), \tag{B.15}$$

which states that if there is any scattering at all, there must be some scattering in the forward direction.

Consider the attenuation (through scattering, nuclear reactions, etc.) of a beam of particles as it passes through a slab of material. An absorbing medium can be regarded as having a complex index of refraction, the imaginary part of which is associated with attenuation. We thus write

$$n = n_{\text{sc}} + n_{\text{abs}} = \left[1 + \frac{2\pi N}{k^2}\, \mathfrak{f}_{\text{sc}}(\omega, 0)\right] + \mathrm{i}\frac{2\pi N}{k^2}\, \mathfrak{f}_{\text{abs}}(\omega, 0). \tag{B.16}$$

Similar to (B.12), the amplitude in this case reads

$$A(z) = A_0 \mathrm{e}^{\mathrm{i}nkz} = A_0 \mathrm{e}^{\mathrm{i}n_{\text{sc}}kz}\mathrm{e}^{-n_{\text{abs}}kz}. \tag{B.17}$$

The normalized beam intensity is

$$\frac{\mathsf{I}}{\mathsf{I}_0} = \mathrm{e}^{-2n_{\text{abs}}kz} = \mathrm{e}^{-4\pi N \mathfrak{f}_{\text{abs}}(\omega, 0)/k} \equiv \mathrm{e}^{-zN\sigma_{\text{tot}}} \tag{B.18}$$

so that

$$\sigma_{\text{tot}} = \frac{4\pi}{k}\, \mathfrak{f}_{\text{abs}}(\omega, 0) = \frac{4\pi}{k}\, \text{Im}\, \mathfrak{f}(\omega, 0), \tag{B.19}$$

which is the optical theorem (B.15). Therefore, the imaginary part of n describes the attenuation of the beam as governed by the total cross-section — not by the absorption cross-section. The real part of n gives the phase shift of the wave relative to its propagation in free space.

C Watson's Theorem and the Decay Amplitudes $K^0, \bar{K}^0 \to 2\pi$

The *unitarity condition*

$$SS^\dagger = 1 \tag{C.1}$$

is imposed on the *scattering matrix* S to ensure probability conservation. In matrix form this condition reads

$$(SS^\dagger)_{fi} = \sum_n S_{fn} S_{in}^* = \delta_{fi}, \tag{C.2}$$

where $\sum_f \delta_{fi} = 1$ and n runs over all possible intermediate states. The diagonal elements satisfy

$$\sum_n |S_{ni}|^2 = 1, \tag{C.3}$$

which expresses the fact that every intermediate state must disintigrate into one of the available final states, i.e., that the sum of the transition probabilities from a given initial state to all final states is unity.

The *transition matrix* T_{fi} is defined by

$$S_{fi} = \delta_{fi} - i(2\pi)^4 \delta^{(4)}(p_f - p_i) T_{fi}. \tag{C.4}$$

Substituting (C.4) in (C.2) yields

$$i(T_{fi} - T_{if}^*) = (2\pi)^4 \sum_n \delta^{(4)}(p_i - p_n) T_{fn} T_{in}^*. \tag{C.5}$$

If we take $f = i$ (forward elastic scattering), this expression leads to the optical theorem (B.15).

The unitarity condition (C.1) should hold true whether or not weak interactions are included. Let δS denote the small change in S due to the weak interaction: $S = S_0 + \delta S$, with

$$\delta S = -i(2\pi)^4 \delta^{(4)}(p_f - p_i) T. \tag{C.6}$$

Using (C.1) we have, to first-order,

$$S^\dagger S = (S_0^\dagger + \delta S^\dagger)(S_0 + \delta S) = \underbrace{S_0^\dagger S_0}_{1} + \delta S^\dagger S_0 + S_0^\dagger \delta S + \underbrace{\delta S^\dagger \delta S}_{\text{neglect}} = 1,$$

i.e.

$$\delta S^\dagger S_0 + S_0^\dagger \delta S = 0. \tag{C.7}$$

Expressions (C.6) and (C.7) give

$$T^\dagger S_0 - S_0^\dagger T = 0, \tag{C.8}$$

or in terms of transition amplitudes

$$\langle f \mid T^\dagger S_0 \mid i \rangle = \langle f \mid S_0^\dagger T \mid i \rangle. \tag{C.9}$$

Since $\langle f \mid \hat{O} \mid i \rangle = \langle i \mid \hat{O}^\dagger \mid f \rangle^*$, it follows that

$$\langle i \mid S_0^\dagger T \mid f \rangle^* = \langle f \mid S_0^\dagger T \mid i \rangle. \tag{C.10}$$

If we ignore the effect of strong and electromagnetic interactions, we can set $S_0 = 1$ in (C10). From the resulting relation we infer that T is hermitian:

$$\langle i \mid T \mid f \rangle^* = \langle f \mid T \mid i \rangle \longrightarrow T = T^\dagger. \tag{C.11}$$

Now, recall that CPT invariance implies (see (3.10))

$$\langle i \mid T \mid f \rangle = \langle \bar{f}' \mid T \mid \bar{i}' \rangle, \quad CPT. \tag{C.12}$$

Combining this result with (C.11), which is based on the unitarity condition (C.1), we obtain

$$\langle f \mid T \mid i \rangle = \langle \bar{f}' \mid T \mid \bar{i}' \rangle^*, \quad CPT, \tag{C.13}$$

in agreement with (3.57).

In vacuum, the initial state $|K^0\rangle$ would be stable in the absence of weak interactions:

$$S_0|i\rangle = |i\rangle, \quad \langle i|S_0^\dagger = \langle i|. \tag{C.14}$$

Since kaons are not stable, we must take into account strong interaction effects among their decay products, as discussed in Sect. 3.4. In the decay $K^0 \to 2\pi$, for example, the two final state pions undergo elastic scattering. Consequently, the phase of their wave function changes by an angle (called the *phase shift*), which is conventionally denoted by $2\delta_l$, where $\delta_{l=0}$ and $\delta_{l=2}$ are S-wave phase shifts for $|0, 0\rangle$ and $|2, 0\rangle$, respectively. We thus write

$$S_0|f\rangle = \mathrm{e}^{\mathrm{i}2\delta_l}|f\rangle, \quad \langle f|S_0^\dagger = \mathrm{e}^{-\mathrm{i}2\delta_l}\langle f|, \tag{C.15}$$

which ensures that S_0 is diagonal. Based on (C.14) and (C.15), equation (C.10) becomes

$$\langle i \mid T \mid f \rangle^* = \mathrm{e}^{-\mathrm{i}2\delta_l}\langle f \mid T \mid i \rangle. \tag{C.16}$$

Expression (C.12) now reads

$$\langle \bar{f}' \mid T \mid \bar{i}' \rangle = \mathrm{e}^{\mathrm{i}2\delta_l} \langle f \mid T \mid i \rangle^*, \tag{C.17}$$

which is Watson's theorem (3.58) (see [14c]). Note that, due to the strong interaction phase shift in the two-pion wave function, the particle and antiparticle decay amplitudes are not complex conjugates of each other.

Using (3.63) and (3.55), the transition amplitudes $A(K^0 \to 2\pi)$ and $A(\bar{K}^0 \to 2\pi)$ can be written as

$$A(K^0, \bar{K}^0 \to \pi^0\pi^0) = \sqrt{\frac{2}{3}}|A_2|\mathrm{e}^{\pm\mathrm{i}\phi_2}\mathrm{e}^{\mathrm{i}\delta_2} - \sqrt{\frac{1}{3}}|A_0|\mathrm{e}^{\pm\mathrm{i}\phi_0}\mathrm{e}^{\mathrm{i}\delta_0},$$

$$A(K^0, \bar{K}^0 \to \pi^+\pi^-) = \sqrt{\frac{1}{3}}|A_2|\mathrm{e}^{\pm\mathrm{i}\phi_2}\mathrm{e}^{\mathrm{i}\delta_2} + \sqrt{\frac{2}{3}}|A_0|\mathrm{e}^{\pm\mathrm{i}\phi_0}\mathrm{e}^{\mathrm{i}\delta_0}, \tag{C.18}$$

where the upper (lower) signs refer to K^0 ($\bar{K}^0$) and $A_\alpha = |A_\alpha|\mathrm{e}^{\mathrm{i}\phi_\alpha}$ with $\alpha = 0$ or 2. The expressions for the corresponding transition rates, which include the interference between the $I = 0$ and $I = 2$ modes, are

$$\Gamma(K^0, \bar{K}^0 \to \pi^0\pi^0) = \frac{1}{3}|A_0|^2 + \frac{2}{3}|A_2|^2$$
$$- \frac{2\sqrt{2}}{3}|A_0||A_2|\cos\left[\pm(\phi_0 - \phi_2) + (\delta_0 - \delta_2)\right],$$

$$\Gamma(K^0, \bar{K}^0 \to \pi^+\pi^-) = \frac{2}{3}|A_0|^2 + \frac{1}{3}|A_2|^2$$
$$+ \frac{2\sqrt{2}}{3}|A_0||A_2|\cos\left[\pm(\phi_0 - \phi_2) + (\delta_0 - \delta_2)\right]. \tag{C.19}$$

From (C.19) we see that:

(a) $\Gamma(K^0 \to 2\pi; I = 0 \text{ or } 2) = \Gamma(\bar{K}^0 \to 2\pi; I = 0 \text{ or } 2) \propto |A_0|^2$ or $|A_2|^2$;

(b) $\Gamma(K^0 \to \pi^0\pi^0 \text{ or } \pi^+\pi^-) = \Gamma(\bar{K}^0 \to \pi^0\pi^0 \text{ or } \pi^+\pi^-)$ only if $\delta_0 = \delta_2$;

(c) $\Gamma(K^0, \bar{K}^0 \to \pi^0\pi^0) + \Gamma(K^0, \bar{K}^0 \to \pi^+\pi^-) = |A_0|^2 + |A_2|^2$.

The difference in $\Gamma(K^0 \to \pi^0\pi^0 \text{ or } \pi^+\pi^-)$ and $\Gamma(\bar{K}^0 \to \pi^0\pi^0 \text{ or } \pi^+\pi^-)$ is due to $\phi_0 - \phi_2$, and is therefore associated with the parameter ε' (see (3.75)), which determines the magnitude of direct CP violation. For reasons unrelated to CP invariance, $\mathrm{Re}\, A_0 \approx 20 \times \mathrm{Re}\, A_2$, as shown in (3.70), (3.71). The intrinsic scale for direct CP violation in weak decays of neutral kaons is thus $\mathcal{O}(10^{-5} - 10^{-4})$ (see (5.7) and (5.11)).

D Time Reversal and *CPT* Violation

To illustrate the effect of time reversal on a quantum-mechanical state, consider the Schrödinger equation

$$i\frac{d}{dt}\Psi(x,t) = \hat{H}\Psi(x,t). \tag{D.1}$$

If the hamiltonian $\hat{H}$ is invariant under time reversal, the new wavefunction obtained under this transformation must also satisfy (D.1). As can be readily verified by complex conjugation of the above equation, the function $\Psi^*(x,-t)$ is a solution of (D.1) provided $\hat{H} = \hat{H}^*$. We thus conclude that time reversal has something to do with complex conjugation, and that a quantum-mechanical state is invariant under time reversal if its hamiltonian is real.

In the above elementary example the time-reversed state was obtained simply by complex conjugation, which is not the case in general. For wavefunctions in momentum space, time reversal implies $\Phi(p) \to \Phi^*(-p)$, that is, in addition to complex conjugation there is also the transformation $\Phi(p) \to \Phi(-p)$.

We can generalize our discussion by introducing the time-independent operator $\hat{T}$, which has the effect of creating a new wavefunction and a new hamiltonian (see, e.g., [45] and [97]):

$$\left.\begin{array}{l} \hat{T}\Psi = \Psi_T \\ \hat{T}\hat{H}\hat{T}^{-1} = \hat{H}_T \end{array}\right\} \hat{T}\hat{T}^{-1} = 1. \tag{D.2}$$

We require that Ψ_T satisfy the Schrödinger equation with $t \to -t$ and hamiltonian $\hat{H}_T$, as does Ψ with hamiltonian $\hat{H}$. If $\hat{H}$ is invariant under time reversal ($\hat{H} = \hat{H}_T$), the operator $\hat{T}$ commutes with the hamiltonian: $\hat{T}\hat{H} = \hat{H}\hat{T}$.

If (D.1) is multiplied by $\hat{T}$ from the left, we obtain

$$\hat{T}i\hat{T}^{-1}\frac{d}{dt}\hat{T}\Psi = \hat{T}\hat{H}\hat{T}^{-1}\hat{T}\Psi. \tag{D.3}$$

By comparing (D.3) with the Schrödinger equation for the time-reversed wavefunction Ψ_T

$$i\frac{d\Psi_T}{d(-t)} = -i\frac{d\Psi_T}{dt} = \hat{H}_T\Psi_T \tag{D.4}$$

we infer that

$$\hat{T}i\hat{T}^{-1} = -i. \tag{D.5}$$

This is a special case of the property

$$\hat{K}(c\Psi) = c^*\Psi^* \to \hat{K} = \hat{K}^{-1} \quad \text{or} \quad \hat{K}^2 = 1, \tag{D.6}$$

which defines an *antilinear* operator. For comparison, the corresponding equation for a *linear* operator is

$$\hat{O}(c\Psi) = c\hat{O}\Psi. \tag{D.7}$$

Therefore, unlike other transformation operators in quantum mechanics, $\hat{T}$ is not a linear operator. Instead, $\hat{T}$ can be expressed as the product of two operators $\hat{U}$ and $\hat{K}$, where $\hat{U}$ is a *unitary* operator ($\hat{U}^\dagger\hat{U} = \hat{U}^{-1}\hat{U} = 1$) and $\hat{K}$ changes a quantity into its complex conjugate. Such an operator is called antilinear-unitary, or *antiunitary*.

To show that

$$\hat{T} = \hat{U}\hat{K}, \tag{D.8}$$

we substitute (D.8) in (D.4) and obtain

$$-i\frac{d\hat{U}\Psi^*}{dt} = \hat{U}(\hat{K}\hat{H}\hat{K}^{-1})\hat{U}^{-1}\hat{U}\Psi^* = \hat{U}\hat{H}^*\hat{U}^{-1}\hat{U}\Psi^*, \tag{D.9}$$

since $(\hat{U}\hat{K})^{-1} = \hat{K}^{-1}\hat{U}^{-1}$ and $\hat{H}^*\Psi^* = \hat{K}(\hat{H}\Psi) = \hat{K}\hat{H}\hat{K}^{-1}\hat{K}\Psi$. Taking the complex conjugate of (D.9) and factoring out the time-independent operator $\hat{U}$, yields the unreversed Schrödinger equation (D.1).

According to (D.2) and (D.8),

$$\Psi_T = \hat{T}\Psi = \hat{U}\Psi^*, \tag{D.10}$$

which we use to express the inner product of two time-reversed states as

$$\begin{aligned}
\langle A_T \mid B_T \rangle &\equiv \int dV (\Psi_{A_T})^\dagger \Psi_{B_T} = \int dV (\hat{U}\Psi_A^*)^\dagger (\hat{U}\Psi_B^*) \\
&= \int dV (\Psi_A^*)^\dagger \hat{U}^\dagger \hat{U} \Psi_B^* = \left\{ \int dV \Psi_A^\dagger \Psi_B \right\}^* \\
&= \langle A \mid B \rangle^* = \langle B \mid A \rangle.
\end{aligned} \tag{D.11}$$

Taking the point of view that the operator $\hat{T}$ is to act on Dirac's kets, we write the time-reversed states

$$|A_T\rangle \equiv \hat{T}|A\rangle, \quad |B_T\rangle \equiv \hat{T}|B\rangle.$$

Let us further define

$$|C\rangle \equiv \hat{O}^\dagger |B\rangle \leftrightarrow \langle C| \equiv \langle B|\hat{O},$$

where $\hat{O}$ is a linear operator (see (D.7)). Now,

$$\begin{aligned}
\langle B \mid \hat{O} \mid A \rangle &= \langle C \mid A \rangle = \langle A_T \mid C_T \rangle = \langle A_T \mid \hat{T}\hat{O}^\dagger \mid B \rangle \\
&= \langle A_T \mid \hat{T}\hat{O}^\dagger\hat{T}^{-1}\hat{T} \mid B \rangle = \langle A_T \mid \hat{T}\hat{O}^\dagger\hat{T}^{-1} \mid B_T \rangle,
\end{aligned}$$

i.e.

$$\langle B \mid \hat{O} \mid A \rangle = \langle A_T \mid \hat{T}\hat{O}^\dagger\hat{T}^{-1} \mid B_T \rangle. \tag{D.12}$$

The time evolution of state vectors in the Schrödinger representation can be expressed as

$$|\Psi(t)\rangle = e^{-i\hat{H}(t-t')}|\Psi(t')\rangle. \tag{D.13}$$

Indeed, if this expression is differentiated with respect to t, we obtain

$$i \frac{d}{dt}|\Psi(t)\rangle = \hat{H}e^{-i\hat{H}(t-t')}|\Psi(t')\rangle = \hat{H}|\Psi(t)\rangle,$$

which is Schrödinger's equation for $|\Psi(t)\rangle$, provided $\hat{H}$ is the total hamiltonian.

Setting $A \equiv \bar{K}^0$, $B \equiv K^0$ and $\hat{O} \equiv e^{-i\hat{H}t}$ in (D.12) gives[44]

$$\langle K^0 \mid e^{-i\hat{H}t} \mid \bar{K}^0 \rangle = \langle \bar{K}^0 \mid e^{-i\hat{H}_T t} \mid K^0 \rangle, \tag{D.14}$$

where

$$\hat{T}e^{i\hat{H}t}\hat{T}^{-1} \equiv e^{-i\hat{H}_T t}. \tag{D.15}$$

Consider the possibility of *CPT* violation in the K^0 system. Based on (3.24), we define

$$\epsilon_1 \equiv \epsilon + \bar{\epsilon}, \quad \epsilon_2 \equiv \epsilon - \bar{\epsilon} \leftrightarrow \epsilon_1 - \epsilon_2 = 2\bar{\epsilon} \tag{D.16}$$

and use equations (3.6) to write the time evolution of an initially pure K^0 or $\bar{K}^0$ state as

$$|K^0(t)\rangle = \frac{1}{\sqrt{2}}\left[(1 - \epsilon + \bar{\epsilon})|K_S^0\rangle e^{-i\mathcal{M}_S t} + (1 - \epsilon - \bar{\epsilon})|K_L^0\rangle e^{-i\mathcal{M}_L t}\right],$$
$$|\bar{K}^0(t)\rangle = \frac{1}{\sqrt{2}}\left[(1 + \epsilon - \bar{\epsilon})|K_S^0\rangle e^{-i\mathcal{M}_S t} - (1 + \epsilon + \bar{\epsilon})|K_L^0\rangle e^{-i\mathcal{M}_L t}\right]. \tag{D.17}$$

Substituting in (D.17) expressions (3.5) for $|K_S^0\rangle$ and $|K_L^0\rangle$:

$$|K_S^0\rangle = \frac{1}{\sqrt{2}}\left[(1 + \epsilon + \bar{\epsilon})|K^0\rangle + (1 - \epsilon - \bar{\epsilon})|\bar{K}^0\rangle\right],$$
$$|K_L^0\rangle = \frac{1}{\sqrt{2}}\left[(1 + \epsilon - \bar{\epsilon})|K^0\rangle - (1 - \epsilon + \bar{\epsilon})|\bar{K}^0\rangle\right], \tag{D.18}$$

we obtain

$$|K^0(t)\rangle = [f_+(t) + 2\bar{\epsilon}f_-(t)]|K^0\rangle + (1 - 2\epsilon)f_-(t)|\bar{K}^0\rangle,$$
$$|\bar{K}^0(t)\rangle = [f_+(t) - 2\bar{\epsilon}f_-(t)]|\bar{K}^0\rangle + (1 + 2\epsilon)f_-(t)|K^0\rangle \tag{D.19}$$

with

$$f_\pm = \frac{1}{2}\left[e^{-i\mathcal{M}_S t} \pm e^{-i\mathcal{M}_L t}\right]. \tag{D.20}$$

Hence,

$$\Gamma(K^0 \to \bar{K}^0) = |f_-(t)|^2(1 - 4\text{Re } \epsilon),$$
$$\Gamma(\bar{K}^0 \to K^0) = |f_-(t)|^2(1 + 4\text{Re } \epsilon). \tag{D.21}$$

---

[44] This result is valid only in the two-dimensional Hilbert space of the K^0 system, with $\hat{H}$ defined by (2.21).

resulting in the time-reversal asymmetry (see (6.19) and (6.20))

$$\mathcal{A}_T(t) \equiv \frac{\Gamma(\bar{K}^0 \to K^0) - \Gamma(K^0 \to \bar{K}^0)}{\Gamma(\bar{K}^0 \to K^0) + \Gamma(K^0 \to \bar{K}^0)} = 4\mathrm{Re}\ \epsilon. \tag{D.22}$$

Note that the transition rates (D.21), which were obtained without any reference to the method of detecting the two K^0 states, do not depend on the CPT violating, T invariant parameter $\bar{\epsilon}$ (see [44]). In contrast,

$$\Gamma(K^0 \to K^0) \approx |f_+(t)|^2 + 2\left[f_+ f_-^* \bar{\epsilon}^* + f_- f_+^* \bar{\epsilon}\right],$$

$$\Gamma(\bar{K}^0 \to \bar{K}^0) \approx |f_+(t)|^2 - 2\left[f_+ f_-^* \bar{\epsilon}^* + f_- f_+^* \bar{\epsilon}\right], \tag{D.23}$$

which yields

$$\mathcal{A}_{CPT}(t) \equiv \frac{\Gamma(\bar{K}^0 \to \bar{K}^0) - \Gamma(K^0 \to K^0)}{\Gamma(\bar{K}^0 \to \bar{K}^0) + \Gamma(K^0 \to K^0)} = -\,4\mathrm{Re}\ \bar{\epsilon}$$

for $t \gg \tau_s$. \hfill (D.24)

E Transformation Properties of Dirac Fields Under C, P and T

Using Maxwell's equations $\partial_\mu F^{\mu\nu} = j^\nu$, where $F^{\mu\nu} \equiv \partial^\mu A^\nu - \partial^\nu A^\mu$ is the electromegnatic field tensor, and the transformation properties of $\partial_\mu \equiv \partial/\partial x^\mu$ and j^μ, one can determine how the electromagnetic potential $A^\mu = (A^0, \boldsymbol{A})$ transforms under parity, P, charge conjugation, C, and time reversal, T. Similarly, the transformation properties of Dirac spinors follow from his relativistic equation

$$\{i\gamma^\mu \partial_\mu - m\}\psi(x) = 0 \quad \text{or}$$

$$i\frac{\partial \psi(x)}{\partial t} = H\psi(x) = \left\{-i\alpha^j \frac{\partial}{\partial x^j} + \beta m\right\}\psi(x). \tag{E.1}$$

In the Dirac–Pauli representation,

$$\alpha^j = \begin{pmatrix} 0 & \sigma^j \\ \sigma^j & 0 \end{pmatrix}, \quad \beta = \gamma^0 = \begin{pmatrix} \mathbb{1} & 0 \\ 0 & -\mathbb{1} \end{pmatrix}, \quad \gamma^j = \begin{pmatrix} 0 & \sigma^j \\ -\sigma^j & 0 \end{pmatrix}, \tag{E.2a}$$

where σ^j are the Pauli spin matrices

$$\sigma^1 = \begin{pmatrix} 0 & 1 \\ 1 & 0 \end{pmatrix}, \sigma^2 = \begin{pmatrix} 0 & -i \\ i & 0 \end{pmatrix}, \sigma^3 = \begin{pmatrix} 1 & 0 \\ 0 & -1 \end{pmatrix} \tag{E.2b}$$

and $\mathbb{1}$ is the unit 2×2 matrix.

The results are shown in Table E.1. The transpose matrix $\overline{\psi}^{\mathrm{T}} = \gamma^0 \psi^* = \gamma^0 \hat{K}\psi$ is expressed in terms of the the complex-conjugation operator $\hat{K}$ defined in (D.6). In the representation (E.2b), the spinor transformation matrices read $\hat{C} = i\gamma^2\gamma^0$ and $\hat{T} = i\gamma^2\gamma_5$, where $\gamma_5 \equiv i\gamma^0\gamma^1\gamma^2\gamma^3$.

Table E.1.

$x^\mu = (t, \boldsymbol{x})$	∂_μ	j^μ	A_μ	ψ	$\overline{\psi}$
P	$(t, -\boldsymbol{x})$	∂^μ	j_μ	A^μ	$\gamma^0 \psi$	$\overline{\psi}\gamma^0$
C	$(t, \boldsymbol{x})$	∂_μ	$-j^\mu$	$-A_\mu$	$\hat{C}\,\overline{\psi}^{\mathrm{T}}$	$-\psi^{\mathrm{T}}\hat{C}^{-1}$
T	$(-t, \boldsymbol{x})$	$-\partial^\mu$	j_μ	A^μ	$\hat{T}\,\overline{\psi}^{\mathrm{T}}$	$\psi^{\mathrm{T}}\hat{T}^{-1}$

The *gauge* (or phase) invariance of the Dirac theory is ensured through the replacement $i\partial_\mu \to i\partial_\mu + eA_\mu$: the transformation $\psi(x) \to \mathrm{e}^{\mathrm{i}\alpha(x)}\psi(x)$ introduces a vector (gauge) field A_μ, which couples to the electron via $\mathfrak{L}_{\mathrm{int}} \equiv -j^\mu A_\mu = e\overline{\psi}\gamma^\mu\psi A_\mu$.

In the following we will outline a formal theory of the discrete symmetry operations P, C and T. Although most of the ensuing discussion can be carried out without recourse to second quantization, the invariance under charge conjugation is an important exception. We thus seek to establish a formalism that would allow us to include also the more elementary symmetry operations, such as space reflections, in theories with second quantization.

Parity Operation

We begin with the *parity operation*, or space inversion, $P\colon \boldsymbol{x} \to \boldsymbol{x}' = -\boldsymbol{x}$, $t \to t' = t$. This is a subtle concept whose true *raison d'être* lies in elementary particle physics.

If we compare (E.1) with the parity-transformed Dirac equation

$$\mathrm{i}\frac{\partial\psi(t, -\boldsymbol{x})}{\partial t} = H'\psi(t, -\boldsymbol{x}) = \left\{ \mathrm{i}\alpha^j \frac{\partial}{\partial x^j} + \beta m \right\}\psi(t, -\boldsymbol{x}) \tag{E.3}$$

and note that $\gamma^0\alpha^j\gamma^0 = -\alpha^j$, $(\gamma^0)^2 = 1$, we see that the hamiltonian has the following symmetry property:

$$H' = \gamma^0 H \gamma^0. \tag{E.4}$$

Equation (E.3) can thus be written as

$$\mathrm{i}\frac{\partial\psi(t, -\boldsymbol{x})}{\partial t} = \gamma^0 H \gamma^0 \,\psi(t, -\boldsymbol{x}). \tag{E.5}$$

We now define a new function

$$\psi_p(t, \boldsymbol{x}) \equiv \gamma^0\psi(t, -\boldsymbol{x}) = \hat{P}\psi(t, -\boldsymbol{x}) \tag{E.6a}$$

and multiply (E.5) by γ^0 from the left, with the result

$$\mathrm{i}\frac{\partial\psi_p(t, \boldsymbol{x})}{\partial t} = H\psi_p(t, \boldsymbol{x}).$$

The above equation is formally identical with (E.1). The Dirac theory is, therefore, invariant under space reflections provided the parity-transformed spinor is defined according to (E.6a). The operator $\hat{P}$, which connects the original and parity-transformed spinors, is *linear-unitary*:

$$\hat{P} = \hat{P}^{-1} \quad \text{i.e.} \quad \hat{P}^2 = 1. \tag{E.7}$$

This expresses the fact that the relation between the two states is reciprocal.

The adjoint spinor has the following transformation property

$$\overline{\psi}_p(t, \boldsymbol{x}) = \psi_p^\dagger(t, \boldsymbol{x})\gamma^0 = \left[\gamma^0 \psi(t, -\boldsymbol{x})\right]^\dagger \gamma^0 = \overline{\psi}(t, -\boldsymbol{x})\gamma^0. \tag{E.6b}$$

If $\psi(x)$ is regarded as a field operator, it can be expanded in terms of plane waves:

$$\psi(t, \boldsymbol{x}) = \int \frac{\mathrm{d}^3 \boldsymbol{p}}{(2\pi)^{3/2}} \sqrt{\frac{m}{E}}$$
$$\times \sum_{s = \pm 1/2} \left[\boldsymbol{d}_s^\dagger(p)v_s(p)\mathrm{e}^{\mathrm{i}px} + \boldsymbol{b}_s(p)u_s(p)\mathrm{e}^{-\mathrm{i}px}\right] \tag{E.8}$$

with the expansion coefficients as operators. The operator $\boldsymbol{d}_s^\dagger(p)$ creates a positron with the z component of spin $s = \pm 1/2$: $\boldsymbol{d}_s^\dagger(p)|0\rangle = |\mathrm{e}^+(\boldsymbol{p}, s)\rangle$, whereas $\boldsymbol{b}_s(p)$ annihilates an electron.

The quantum fields $\psi_p(t, \boldsymbol{x})$ and $\psi(t, \boldsymbol{x})$ satisfy the same equation of motion. Furthermore, it can be readily verified that they obey the same anticommutation rules

$$\{\psi_\alpha(t, \boldsymbol{x}), \psi_\beta(t, \boldsymbol{y})\} = \{\overline{\psi}_\alpha(t, \boldsymbol{x}), \overline{\psi}_\beta(t, \boldsymbol{y})\} = 0,$$
$$\{\psi_\alpha(t, \boldsymbol{x}), \overline{\psi}_\beta(t, \boldsymbol{y})\} = (\gamma^0)_{\alpha\beta}\, \delta^{(3)}(\boldsymbol{x} - \boldsymbol{y}). \tag{E.9a}$$

Indeed,

$$\{\psi_\alpha^p(t, \boldsymbol{x}), \overline{\psi}_\beta^p(t, \boldsymbol{y})\} = (\gamma^0)_{\alpha\varrho}(\gamma^0)_{\sigma\beta}\{\psi_\varrho(t, -\boldsymbol{x}), \overline{\psi}_\sigma(t, -\boldsymbol{y})\}$$
$$= (\gamma^0)_{\alpha\beta}\, \delta^{(3)}(\boldsymbol{x} - \boldsymbol{y}), \tag{E.9b}$$

and similarly

$$\{\psi_\alpha^p(t, \boldsymbol{x}), \psi_\beta^p(t, \boldsymbol{y})\} = \{\overline{\psi}_\alpha^p(t, \boldsymbol{x}), \overline{\psi}_\beta^p(t, \boldsymbol{y})\} = 0. \tag{E.9c}$$

We thus expect the field operators $\psi_p(t, \boldsymbol{x})$ and $\psi(t, \boldsymbol{x})$ to be related by means of a unitary transformation:

$$\boldsymbol{U}(P)\psi(t, \boldsymbol{x})\boldsymbol{U}^\dagger(P) \overset{?}{=} \psi_p(t, \boldsymbol{x}) \equiv \gamma^0 \psi(t, -\boldsymbol{x}). \tag{E.10}$$

When the Fourier expansion (E.8) is substituted into this relation, it reads

$$\boldsymbol{U}(P)\psi(t, \boldsymbol{x})\boldsymbol{U}^\dagger(P) = \int \frac{\mathrm{d}^3 \boldsymbol{p}}{(2\pi)^{3/2}} \sqrt{\frac{m}{E}} \sum_s \left[\boldsymbol{U}(P)\boldsymbol{d}_s^\dagger(p)\boldsymbol{U}^\dagger(P)v_s(p)\mathrm{e}^{\mathrm{i}px}\right.$$
$$\left. + \boldsymbol{U}(P)\boldsymbol{b}_s(p)\boldsymbol{U}^\dagger(P)u_s(p)\mathrm{e}^{-\mathrm{i}px}\right]$$
$$\equiv \int \frac{\mathrm{d}^3 \boldsymbol{p}}{(2\pi)^{3/2}} \sqrt{\frac{m}{E}} \sum_s \left[\boldsymbol{d}_s^\dagger(-\boldsymbol{p})\gamma^0 v_s(p')\mathrm{e}^{\mathrm{i}px}\right.$$
$$\left. + \boldsymbol{b}_s(-\boldsymbol{p})\gamma^0 u_s(p')\mathrm{e}^{-\mathrm{i}px}\right], \tag{E.11}$$

with $p' \equiv (E, -\boldsymbol{p})$. By using the explicit form of the Dirac spinors

$$u_s = \sqrt{\frac{E+m}{2m}} \left(\begin{array}{c} \chi_s \\ \frac{\boldsymbol{\sigma}\cdot\boldsymbol{p}}{E+m} \chi_s \end{array} \right)_{E>0},$$

$$v_s = \sqrt{\frac{E+m}{2m}} \left(\begin{array}{c} \frac{-\boldsymbol{\sigma}\cdot\boldsymbol{p}}{|E|+m} \chi_s \\ \chi_s \end{array} \right)_{E<0}, \tag{E.12a}$$

where

$$\chi_{s=+1/2} = \left(\begin{array}{c} 1 \\ 0 \end{array} \right), \quad \chi_{s=-1/2} = \left(\begin{array}{c} 0 \\ 1 \end{array} \right), \tag{E.12b}$$

the following symmetry properties of u_s and v_s can be easily derived

$$\gamma^0 u_s(p') = u_s(p), \quad \gamma^0 v_s(p') = -v_s(p). \tag{E.13}$$

From (E.11) and (E.13) we infer that

$$\boldsymbol{U}(P)\boldsymbol{b}_s^\dagger(\boldsymbol{p})\boldsymbol{U}^\dagger(P) = \boldsymbol{b}_s^\dagger(-\boldsymbol{p}),$$

$$\boldsymbol{U}(P)\boldsymbol{d}_s^\dagger(\boldsymbol{p})\boldsymbol{U}^\dagger(P) = -\boldsymbol{d}_s^\dagger(-\boldsymbol{p}), \tag{E.14}$$

where the first expression is the hermitian conjugate of $\boldsymbol{U}(P)\boldsymbol{b}_s(\boldsymbol{p})\boldsymbol{U}^\dagger(P) = \boldsymbol{b}_s(-\boldsymbol{p})$ (recall that $\boldsymbol{U}^\dagger\boldsymbol{U} = \boldsymbol{U}^{-1}\boldsymbol{U} = 1$ for a unitary operator $\boldsymbol{U}$).

If we further assume that $\boldsymbol{U}(P)|0\rangle = |0\rangle$, i.e., that the vacuum is invariant under P, then

$$\boldsymbol{U}(P)|e^-(\boldsymbol{p}, s)\rangle = |e^-(-\boldsymbol{p}, s)\rangle,$$

$$\boldsymbol{U}(P)|e^+(\boldsymbol{p}, s)\rangle = -|e^+(-\boldsymbol{p}, s)\rangle. \tag{E.15}$$

The unitarity of $\boldsymbol{U}$ ensures that the state vectors are orthonormal:

$$\langle e^-(\boldsymbol{p}, s) \,|\, \boldsymbol{U}^\dagger\boldsymbol{U} \,|\, e^-(\boldsymbol{p}, s)\rangle = \langle e^-(-\boldsymbol{p}, s) \,|\, e^-(-\boldsymbol{p}, s)\rangle. \tag{E.16}$$

Expressions (E.14) define a unitary operator in the space of electron and positron state vectors[45] that satisfies (E.10). Based on the above results we conclude that for every state of free electrons and positrons there exists a parity-transformed state in which all momenta are reversed, but the spins are not affected. To test this prediction, it is necessary to set up a physical situation which cannot revert back into itself under space inversion. By aligning the spins of β-emitting ^{60}Co nuclei in a strong magnetic field and then measuring the relative electron intensities along and against the field, C. S. Wu and her collaborators were able to demonstrate, in 1957, that the left–right symmetry of free Dirac fields is violated in weak interactions.

Of particular interest is the second equality in (E.15). It shows that the *intrinsic* parity of the electron is opposite to that of the positron. This prediction has been verified experimentally, for example, in the decay of positronium.

---

[45] The inversion operators $\boldsymbol{U}(P, C)$ and $\boldsymbol{V}(T)$ act in Hilbert space, whereas $\hat{P}$, $\hat{C}$ and $\hat{T}$ (which are composed of γ matrices) operate in spinor space.

Charge Conjugation

We next consider the operation of *charge conjugation*. The adjoint Dirac field operator

$$\bar{\psi}(x) = \int \frac{\mathrm{d}^3\boldsymbol{p}}{(2\pi)^{3/2}} \sqrt{\frac{m}{E}}$$
$$\times \sum_{s=\pm 1/2} \left[\boldsymbol{b}_s^\dagger(\boldsymbol{p}) \bar{u}_s(p) \mathrm{e}^{\mathrm{i}px} + \boldsymbol{d}_s(\boldsymbol{p}) \bar{v}_s(p) \mathrm{e}^{-\mathrm{i}px} \right] \tag{E.17}$$

annihilates positrons and creates electrons, whereas $\psi(x)$ does exactly the opposite. It is, therefore, plausible to expect that the Dirac equation is invariant under the replacement $\psi(x) \leftrightarrow \bar{\psi}(x)$. To see if this assumption is correct, we take the complex conjugate of (E.1) and obtain

$$[\mathrm{i}(-\gamma^\mu)^\mathrm{T}\partial_\mu - m]\bar{\psi}^\mathrm{T} = 0, \tag{E.18}$$

where we used

$$\gamma^0(\gamma^\mu)^*\gamma^0 = (\gamma^\mu)^\mathrm{T}, \quad \bar{\psi}^\mathrm{T} \equiv (\psi^+\gamma^0)^\mathrm{T} = \gamma^0\psi^*. \tag{E.19}$$

If we multiply (E.18) by $\hat{C}$ from the left and compare the result with (E.1), we find that two equations are equivalent if there exists an operator $\hat{C}$ such that $\hat{C}(-\gamma^\mu)^\mathrm{T} = \gamma^\mu\hat{C}$, i.e.,

$$\hat{C}^{-1}\gamma^\mu\hat{C} = (-\gamma^\mu)^\mathrm{T}. \tag{E.20}$$

A suitable choice for $\hat{C}$ is

$$\hat{C} = \mathrm{i}\gamma^2\gamma^0 \longrightarrow \hat{C} = -\hat{C}^{-1} = -\hat{C}^\dagger. \tag{E.21}$$

Under space inversion, the charge-conjugated spinor $\psi_c(x)$ transforms as

$$\psi_c \equiv \hat{C}\bar{\psi}^\mathrm{T} = \hat{C}(\psi^\dagger\gamma^0)^\mathrm{T} \xrightarrow{\hat{P}} \hat{C}[(\gamma^0\psi)^\dagger\gamma^0]^\mathrm{T}$$
$$= \hat{C}(\bar{\psi}\gamma^0)^\mathrm{T} = \hat{C}\gamma^0\hat{C}^{-1}\hat{C}\bar{\psi}^\mathrm{T} = -\gamma^0\psi_c.$$

Therefore,

$$\psi(x) \xrightarrow{\hat{P}} \gamma^0\psi(x), \quad \psi_c(x) \xrightarrow{\hat{P}} -\gamma^0\psi_c(x), \tag{E.22}$$

which shows that the intrinsic parity of a particle is indeed opposite to that of its antiparticle (cf. (E.15)).

To demonstrate the equivalence of electrons and positrons in the Dirac theory, we look for a unitary operator $\boldsymbol{U}(C)$ that satisfies the following relation among field operators:

$$\boldsymbol{U}(C)\psi(x)\boldsymbol{U}^\dagger(C) \stackrel{?}{=} \psi_c(x) \equiv \hat{C}\bar{\psi}^\mathrm{T}(x) = \hat{C}\left[\psi^\dagger(x)\gamma^0\right]^\mathrm{T}, \tag{E.23}$$

i.e.,

$$U(C)\psi(x)U^\dagger(C)=\int \frac{\mathrm{d}^3 p}{(2\pi)^{3/2}} \sqrt{\frac{m}{E}} \sum_s [U(C)d_s^\dagger(p)U^\dagger(C)v_s(p)\mathrm{e}^{ipx}$$

$$+ \; U(C)b_s(p)U^\dagger(C)u_s(p)\mathrm{e}^{-ipx}]$$

$$\equiv \int \frac{\mathrm{d}^3 p}{(2\pi)^{3/2}} \sqrt{\frac{m}{E}} \sum_s [b_s^\dagger(p)\,\mathrm{i}\gamma^2 u_s^*(p)\mathrm{e}^{ipx}$$

$$+ d_s(p)\,\mathrm{i}\gamma^2 v_s^*(p)\mathrm{e}^{-ipx}]. \tag{E.24}$$

From (E.24) and using

$$\mathrm{i}\gamma^2 u_s^*(p) = v_s(p), \quad \mathrm{i}\gamma^2 v_s^*(p) = u_s(p) \tag{E.25}$$

we obtain

$$U(C)b_s^\dagger(p)U^\dagger(C) = d_s^\dagger(p),$$

$$U(C)d_s^\dagger(p)U^\dagger(C) = b_s^\dagger(p). \tag{E.26}$$

To prove (E.25), consider the following equations for the Dirac spinors

$$(i\gamma^\mu \partial_\mu - m)u = 0, \quad \bar{v}(i\gamma^\mu \partial_\mu + m) = 0. \tag{E.27}$$

If the second equation is transposed and multiplied by $\hat{C}$ from the left, it yields, based on (E.20),

$$(i\gamma^\mu \partial_\mu - m)i\gamma^2 v^* = 0.$$

The relation on the right of (E.25) is obtained by comparing this result with the first equation (E.27). The relation on the left of (E.25) can be proved in a similar manner.

Now, assuming that $U(C)|0\rangle = |0\rangle$, expressions (E.26) lead to

$$U(C)|e^-(p,s)\rangle = |e^+(p,s)\rangle,$$

$$U(C)|e^+(p,s)\rangle = |e^-(p,s)\rangle. \tag{E.28}$$

Under charge conjugation, the roles of creation and annihilation operators are interchanged: $b_s(p) \leftrightarrow b_s^\dagger(p)$ and $d_s(p) \leftrightarrow d_s^\dagger(p)$. This symmetry operation flips the signs of *internal charges*, such as the electric charge, baryon number etc., but spins and momenta are not affected. For example, C turns a left-handed neutrino into a left-handed antineutrino, a state which does not exist. The operation of charge conjugation, therefore, does not transform a particle into its antiparticle; this can be accomplished through the combined *CPT* operation.

Like parity conservation, charge conjugation invariance too is violated in weak interactions. Direct evidence for this violation is provided, for example, by the fact that the positive and negative electrons in the decay $\mu^\pm \to e^\pm \nu \bar{\nu}$ have opposite longitudinal polarization. This effect was first observed in 1957 by measuring the circular polarization of bremsstrahlung photons emitted by e^+ and e^- in the muon decay (the total transmission cross-section for photons propagating through magnetized iron depends on their helicity).

Having defined C conjugation for free Dirac fields, we will now examine the effect of this symmetry transformation on the interaction parts of the lagrangian, $\mathfrak{L}_{\text{int}}$. For two Dirac fields ψ_1 and ψ_2 interacting with a vector field V_μ,

$$\mathfrak{L}_{\text{int}} = g\left(\overline{\psi}_1\gamma^\mu\psi_2 V_\mu + \overline{\psi}_2\gamma^\mu\psi_1 V_\mu^\dagger,\right) \tag{E.29}$$

where g is a real coupling constant. The second term is the hermitian conjugate of the first one, thus ensuring the hermiticity of the lagrangian.

It should be remembered that every quantized field theory that obeys commutation or anticommutation rules must be properly symmetrized or antisymmetrized. Thus all the bilinear forms of the Dirac field must be antisymmetrized. We will encounter shortly an important consequence of this rule.

The adjoint field transforms as

$$\begin{aligned}
\overline{\psi}_c &\equiv \psi_c^\dagger\gamma^0 = (\hat{C}\,\overline{\psi}^{\text{T}})^\dagger\gamma^0 = (\psi^\dagger\gamma^0)^*\hat{C}^\dagger\gamma^0 \\
&= \psi^{\text{T}}\gamma^0\hat{C}^{-1}\gamma^0 = -\psi^{\text{T}}\hat{C}^{-1},
\end{aligned} \tag{E.30}$$

where we used (E.20) and (E.21). Hence,

$$\begin{aligned}
\overline{\psi}_1\gamma^\mu\psi_2 \xrightarrow{\hat{C}} -\psi_1^{\text{T}}\hat{C}^{-1}\gamma^\mu\hat{C}\,\overline{\psi}_2^{\text{T}} &= \psi_1^{\text{T}}(\gamma^\mu)^{\text{T}}\,\overline{\psi}_2^{\text{T}} = -(\overline{\psi}_2\gamma^\mu\psi_1)^{\text{T}} \\
&= -\overline{\psi}_2\gamma^\mu\psi_1
\end{aligned} \tag{E.31}$$

(the superscript "T" can be omitted because $\overline{\psi}\gamma^\mu\psi$ is a number, i.e., a one-by-one "matrix"). The origin of the minus sign in (E.31) is both subtle and important; it is related to the connection between spin and statistics. Since the fermion fields are antisymmetric, a minus sign must be introduced when we move one spinor past another. For the electron current this implies

$$j^\mu(x) \equiv -e\overline{\psi}\gamma^\mu\psi \xrightarrow{\hat{C}} -j^\mu(x). \tag{E.32}$$

The above result was anticipated: C conjugation flips the signs of all charges, including that associated with the current operator. We see that the method of second quantization is indeed essential for a self-consistent formulation of charge conjugation invariance.

From (E.31) it follows that the lagrangian (E.29) is invariant under charge conjugation provided $V_\mu \to -V_\mu^\dagger$, in which case this symmetry transformation merely turns each term in the lagrangian into its hermitian conjugate.

The electromagnetic potential $A_\mu(x)$ and the current $j^\mu(x)$ are related through a simple differential operator (see the beginning of this appendix). Hence they must have the same transformation properties under C, which means that the combination $j^\mu A_\mu$ is invariant under C conjugation. If we associate a photon with the field $A_\mu(x)$, then $C_\gamma = -1$. As shown in Appendix A, for a state with n photons, $C = (-1)^n$. In the decay of positronium, C invariance implies that the $^1S_0$ singlet state decays to two photons and the $^3S_0$ triplet state to three photons.

We conclude our discussion of charge conjugation by showing that ψ_c satisfies the anticommutation rules (E.9a). In terms of field components,

$$\psi_\alpha^c = \hat{C}_{\alpha\varrho}\overline{\psi}_\varrho, \quad \overline{\psi}_\beta^c = -\psi_\sigma(\hat{C}^{-1})_{\alpha\beta},$$

where we dropped the superscript T since it pertains to the complete field operator, not to its components. Now,

$$\{\psi_\alpha^c(t,\boldsymbol{x}), \overline{\psi}_\beta^c(t,\boldsymbol{y})\} = -\hat{C}_{\alpha\varrho}(\hat{C}^{-1})_{\sigma\beta}\{\overline{\psi}_\varrho(t,\boldsymbol{x}), \psi_\sigma(t,\boldsymbol{y})\}$$

$$= -\hat{C}_{\alpha\varrho}(\gamma^0)_{\varrho\sigma}(\hat{C}^{-1})_{\sigma\beta}\,\delta^{(3)}(\boldsymbol{x}-\boldsymbol{y})$$

$$= (\gamma^0)_{\alpha\beta}\,\delta^{(3)}(\boldsymbol{x}-\boldsymbol{y}), \tag{E.33}$$

and similarly

$$\{\psi_\alpha^c(t,\boldsymbol{x}), \psi_\beta^c(t,\boldsymbol{y})\} = \{\overline{\psi}_\alpha^c(t,\boldsymbol{x}), \overline{\psi}_\beta^c(t,\boldsymbol{y})\} = 0. \tag{E.34}$$

Time Reversal

Turning next to the *time-reversal* transformation $(t, \boldsymbol{x}) \to (-t, \boldsymbol{x})$, we will demonstrate that the free Dirac field does not possess a unique time direction, i.e., that it is invariant under this symmetry operation.

As explained in Appendix D, the time-reversal operator $\hat{T}$ is antilinear-unitary (or *antiunitary*):

$$\hat{T} = \text{unitary transformation } (\hat{U}) \times \text{complex conjugation } (\hat{K}). \tag{E.35}$$

Since complex conjugation is involved in time reflections, the Dirac equation for the time-reversed state reads (cf. (E.18))

$$\left\{-i\gamma^0\frac{\partial}{\partial(-t)} + i(-\gamma^j)^{\mathrm{T}}\frac{\partial}{\partial x^j} - m\right\}\overline{\psi}^{\mathrm{T}}(-t,\boldsymbol{x}) = 0. \tag{E.36}$$

From (E.1) and (E.36) it follows that

$$\psi_t(t,\boldsymbol{x}) \equiv \hat{T}\,\overline{\psi}^{\mathrm{T}}(-t,\boldsymbol{x}) \tag{E.37}$$

and

$$\hat{T}^{-1}\gamma^0\hat{T} = \gamma^0 = -\hat{C}^{-1}\gamma^0\hat{C},$$

$$\hat{T}^{-1}\gamma^j\hat{T} = (-\gamma^j)^{\mathrm{T}} = \hat{C}^{-1}\gamma^j\hat{C} \tag{E.38}$$

for the Dirac equation to remain invariant under time reversal. Clearly, the operators $\hat{T}$ and $\hat{C}$ are of similar nature. In the Dirac–Pauli representation,

$$\hat{T} = i\gamma^2\gamma_5 = \gamma^1\gamma^3\gamma^0 \longrightarrow \hat{T} = -\hat{T}^{-1} = -\hat{T}^\dagger, \hat{T}^2 = -1 \tag{E.39}$$

and

$$\hat{T}\,\overline{\psi}^{\mathrm{T}}(-t,\boldsymbol{x}) = \gamma^1\gamma^3\gamma^0\left[\psi^\dagger(-t,\boldsymbol{x})\gamma^0\right]^{\mathrm{T}} = \gamma^1\gamma^3\psi^*(-t,\boldsymbol{x}), \tag{E.40}$$

where

$$\gamma_5 \equiv i\gamma^0\gamma^1\gamma^2\gamma^3 = \begin{pmatrix} 0 & \mathbb{1} \\ \mathbb{1} & 0 \end{pmatrix}. \tag{E.41}$$

The adjoint of the time-reversed spinor is given by

$$\overline{\psi}_t(t, \boldsymbol{x}) = \left[\hat{T}\,\overline{\psi}^{\mathrm{T}}(-t, \boldsymbol{x})\right]^\dagger \gamma^0 = \left[\hat{T}\left(\psi^\dagger(-t, \boldsymbol{x})\gamma^0\right)^{\mathrm{T}}\right]^\dagger \gamma^0$$

$$= \left[\psi^\dagger(-t, \boldsymbol{x})\gamma^0\right]^* \hat{T}^\dagger \gamma^0 = \psi^{\mathrm{T}}(-t, \boldsymbol{x})\hat{T}^{-1}$$

since $\hat{T}^\dagger = \hat{T}^{-1}$ and $\hat{T}^{-1}\gamma^0 = \gamma^0\hat{T}^{-1}$. Therefore,

$$\overline{\psi}(t, \boldsymbol{x}) \xrightarrow{\hat{T}} \overline{\psi}_t(t, \boldsymbol{x}) \equiv \psi^{\mathrm{T}}(-t, \boldsymbol{x})\hat{T}^{-1}. \tag{E.42}$$

Using (E.40) and (E.42), it can be readily shown that the quantization conditions remain invariant under time reversal:

$$\{\psi_\alpha^t(t, \boldsymbol{x}), \overline{\psi}_\beta^{\,t}(t, \boldsymbol{y})\}$$

$$= (\gamma^1\gamma^3)_{\alpha\sigma}(\gamma^0\gamma^3\gamma^1)_{\varrho\beta}\{\psi_\sigma^*(-t, \boldsymbol{x}), \psi_\varrho^{\mathrm{T}}(-t, \boldsymbol{y})\}$$

$$= (\gamma^1\gamma^3)_{\alpha\sigma}(\gamma^0)_{\delta\varrho}(\gamma^0\gamma^3\gamma^1)_{\varrho\beta}\{\psi_\sigma(-t, \boldsymbol{x}), \overline{\psi}_\delta(-t, \boldsymbol{y})\}^*$$

$$= (\gamma^1\gamma^3)_{\alpha\sigma}(\gamma^0)_{\sigma\delta}(\gamma^0)_{\delta\varrho}(\gamma^0\gamma^3\gamma^1)_{\varrho\beta}\,\delta^{(3)}(\boldsymbol{x} - \boldsymbol{y})$$

$$= (\gamma^0)_{\alpha\beta}\,\delta^{(3)}(\boldsymbol{x} - \boldsymbol{y}). \tag{E.43}$$

Similarly,

$$\{\psi_\alpha^t(t, \boldsymbol{x}), \psi_\beta^t(t, \boldsymbol{y})\} = \{\overline{\psi}_\alpha^{\,t}(t, \boldsymbol{x}), \overline{\psi}_\beta^{\,t}(t, \boldsymbol{y})\} = 0. \tag{E.44}$$

In analogy with our treatment of parity and charge conjugation, we seek an antiunitary transformation in Hilbert space that transforms $\psi(x)$ into $\psi_t(x)$. A clue is provided by expression (D.12) from Appendix D:

$$\langle B \mid \hat{O} \mid A \rangle = \langle A_t \mid \hat{O}_t \mid B_t \rangle, \tag{E.45}$$

where

$$\hat{O}_t \equiv \hat{T}\hat{O}^\dagger\hat{T}^{-1} \tag{E.46}$$

and $\hat{O}$ is a linear operator. Note that, due to complex conjugation, the time reversal transformation exchanges "bra" and "ket" vectors. This represents the exchange of the initial and final states in an interaction.

In view of (E.46), we postulate the existence of an antiunitary operator $\boldsymbol{V}(T) \equiv \boldsymbol{U}(T)\hat{K}$ that satisfies

$$\boldsymbol{V}(T)\psi^\dagger(t, \boldsymbol{x})\boldsymbol{V}^{-1}(T) = \psi_t(t, \boldsymbol{x}) \equiv \hat{T}\overline{\psi}^{\mathrm{T}}(-t, \boldsymbol{x}). \tag{E.47}$$

Expressed in terms of field components, equation (E.47) reads

$$\boldsymbol{V}(T)\psi_\alpha^*(t, \boldsymbol{x})\boldsymbol{V}^{-1}(T) = (\gamma^1\gamma^3\gamma^0)_{\alpha\beta}\left[\psi_\varrho^*(-t, \boldsymbol{x})\gamma_{\varrho\beta}^0\right]^{\mathrm{T}}$$

$$= (\gamma^1\gamma^3)_{\alpha\varrho}\psi_\varrho^*(-t, \boldsymbol{x}), \tag{E.48a}$$

i.e.,

$$\boldsymbol{U}(T)\psi_\alpha(t, \boldsymbol{x})\boldsymbol{U}^{-1}(T) = (\gamma^1\gamma^3)_{\alpha\varrho}\psi_\varrho^*(-t, \boldsymbol{x}). \tag{E.48b}$$

By Fourier-transforming this expression into momentum space, we obtain

$$\boldsymbol{U}(T)\psi(x)\boldsymbol{U}^{-1}(T)=\int \frac{d^3\boldsymbol{p}}{(2\pi)^{3/2}} \sqrt{\frac{m}{E}} \sum_s \left[\boldsymbol{U}(T)\boldsymbol{d}_s^\dagger(\boldsymbol{p})\boldsymbol{U}^{-1}(T)v_s(p)\mathrm{e}^{ipx}\right.$$

$$\left.+ \boldsymbol{U}(T)\boldsymbol{b}_s(\boldsymbol{p})\boldsymbol{U}^{-1}(T)u_s(p)\mathrm{e}^{-ipx}\right]$$

$$\equiv\int \frac{d^3\boldsymbol{p}}{(2\pi)^{3/2}} \sqrt{\frac{m}{E}} \sum_s \left[\boldsymbol{d}_s^\dagger(\boldsymbol{p})\gamma^1\gamma^3 v_s^*(p)\mathrm{e}^{i(Et+\boldsymbol{p}\cdot\boldsymbol{x})}\right.$$

$$\left.+ \boldsymbol{b}_s(\boldsymbol{p})\gamma^1\gamma^3 u_s^*(p)\mathrm{e}^{-i(Et+\boldsymbol{p}\cdot\boldsymbol{x})}\right]. \tag{E.49}$$

Now,

$$\gamma^1\gamma^3 u_s^*(p) = i\sigma^2 \begin{pmatrix} \chi_s \\ \dfrac{\boldsymbol{\sigma}^*\cdot\boldsymbol{p}}{E+m}\chi_s \end{pmatrix} \tag{E.50}$$

and

$$i\sigma^2\chi_{s=\pm 1/2} = \begin{cases} -\chi_{s=-1/2} \\ \chi_{s=+1/2} \end{cases}$$

$$i\sigma^2(\boldsymbol{\sigma}^*\cdot\boldsymbol{p})\chi_{s=\pm 1/2} = \begin{cases} -\boldsymbol{\sigma}\cdot(-\boldsymbol{p})\chi_{s=-1/2} \\ \boldsymbol{\sigma}\cdot(-\boldsymbol{p})\chi_{s=+1/2} \end{cases}. \tag{E.51}$$

Equation (E.50) can thus be written as

$$\gamma^1\gamma^3 u^*(p,s) = (-1)^{s+1/2}u(p',-s), \tag{E.52a}$$

with $p' = (E,-\boldsymbol{p})$. Similarly

$$\gamma^1\gamma^3 v^*(p,s) = (-1)^{s+1/2}v(p',-s). \tag{E.52b}$$

When (E.52a) is substituted in (E.49) and $\boldsymbol{p}$ changed to $-\boldsymbol{p}$ in the second integral, we find that

$$\boldsymbol{U}(T)\boldsymbol{b}^\dagger(\boldsymbol{p},s)\boldsymbol{U}^{-1}(T) = (-1)^{s-1/2}\,\boldsymbol{b}^\dagger(-\boldsymbol{p},-s),$$

$$\boldsymbol{U}(T)\boldsymbol{d}^\dagger(\boldsymbol{p},s)\boldsymbol{U}^{-1}(T) = (-1)^{s-1/2}\,\boldsymbol{d}^\dagger(-\boldsymbol{p},-s), \tag{E.53}$$

where the first relation (E.53) is the hermitian conjugate of $\boldsymbol{U}\boldsymbol{b}(\boldsymbol{p},s)\boldsymbol{U}^{-1} = (-1)^{s-1/2}\,\boldsymbol{b}(-\boldsymbol{p},-s)$.

The phase factor in (E.53) implies that the result of two time reversal transformations performed on the Dirac field is the original field multiplied by a minus sign. Indeed, from (E.48a) it follows that

$$\boldsymbol{U}^2(T)\psi(t,\boldsymbol{x})\boldsymbol{U}^{-2}(T) = \boldsymbol{U}(T)\left[\gamma^1\gamma^3\psi(-t,\boldsymbol{x})\right]^*\boldsymbol{U}^{-1}(T)$$

$$= (\gamma^1\gamma^3)(\gamma^1\gamma^3)\psi(t,\boldsymbol{x}) = -\psi(t,\boldsymbol{x}), \tag{E.54}$$

and from (E.53),

$$\boldsymbol{U}^2(T)\boldsymbol{b}^\dagger(\boldsymbol{p},s)\boldsymbol{U}^{-2}(T) = (-1)^{s-1/2}\,\boldsymbol{U}(T)\boldsymbol{b}^\dagger(-\boldsymbol{p},-s)\boldsymbol{U}^{-1}(T)$$
$$= -\,\boldsymbol{b}^\dagger(\boldsymbol{p},s). \tag{E.55}$$

Assuming that $\boldsymbol{U}(T)|\,0\,\rangle = |\,0\,\rangle$, expressions (E.53) yield

$$\boldsymbol{U}(T)\,|\,e^\pm(\boldsymbol{p},s)\rangle = (-1)^{s-1/2}\,|\,e^\pm(-\boldsymbol{p},-s)\rangle. \tag{E.56}$$

This symmetry transformation, therefore, reverses the momentum and spin of an electron (positron) with respect to the original orientation along the z axis. This is to be expected, since $\boldsymbol{p}$ is the time derivative of $\boldsymbol{x}$ and the spin transforms as angular momentum $(\boldsymbol{x}\times\boldsymbol{p})$.

According to (E.45), the expectation values of the observables $\hat{O}$ and $\hat{O}_t$, which are constructed from the field operators ψ and ψ_t, respectively, satisfy

$$\langle A\,|\,\hat{O}\,|\,A\,\rangle = \langle\,A_t\,|\,\hat{O}_t\,|\,A_t\,\rangle, \tag{E.57}$$

where

$$|\,A_t\,\rangle \equiv \boldsymbol{V}(T)|\,A\,\rangle, \quad \boldsymbol{V}^\dagger\boldsymbol{V} = \boldsymbol{V}\boldsymbol{V}^\dagger = 1. \tag{E.58}$$

With the existence of an antiunitary operator that transforms ψ into ψ_t previously established, expression (E.57) demonstrates time-reversal invariance for a free Dirac field.

Transformation Properties of Dirac Bilinears

By virtue of Lorentz invariance, the quark and lepton spinors appear in *bilinear forms* in the lagrangians of quantum field theories. The transformation properties of the Dirac bilinears $\overline{\psi}\psi$ (scalar), $\overline{\psi}\gamma_5\psi$ (pseudoscalar), $\overline{\psi}\gamma^\mu\psi$ (vector), $\overline{\psi}\gamma^\mu\gamma_5\psi$ (axial vector) and $\overline{\psi}\sigma^{\mu\nu}\psi$ (tensor) under the discrete symmetry operations CP, T and CPT are given in Table E.2.

Table E.2.

	$(t,\boldsymbol{x})$	$\overline{\psi}_a\psi_b$	$\overline{\psi}_a\gamma_5\psi_b$	$\overline{\psi}_a\gamma^\mu\psi_b$	$\overline{\psi}_a\gamma^\mu\gamma_5\psi_b$	$\overline{\psi}_a\sigma^{\mu\nu}\psi_b$
CP	$(t,-\boldsymbol{x})$	$\overline{\psi}_b\psi_a$	$-\overline{\psi}_b\gamma_5\psi_a$	$-\overline{\psi}_b\gamma_\mu\psi_a$	$-\overline{\psi}_b\gamma_\mu\gamma_5\psi_a$	$-\overline{\psi}_b\sigma_{\mu\nu}\psi_a$
T	$(-t,\boldsymbol{x})$	$\overline{\psi}_a\psi_b$	$\overline{\psi}_a\gamma_5\psi_b$	$\overline{\psi}_a\gamma_\mu\psi_b$	$\overline{\psi}_a\gamma_\mu\gamma_5\psi_b$	$-\overline{\psi}_a\sigma_{\mu\nu}\psi_b$
CPT	$(-t,-\boldsymbol{x})$	$\overline{\psi}_b\psi_a$	$-\overline{\psi}_b\gamma_5\psi_a$	$-\overline{\psi}_b\gamma^\mu\psi_a$	$-\overline{\psi}_b\gamma^\mu\gamma_5\psi_a$	$\overline{\psi}_b\sigma^{\mu\nu}\psi_a$

The lagrangian of a local field theory must be hermitian[46] and behave either as a scalar or a pseudoscalar under Lorentz transformations. Based on this one can show, referring to Table E.2, that CPT is a good symmetry. For example, a term in the lagrangian $\mathfrak{L}$ that includes only scalars and/or pseudoscalars transforms under CPT as

---

[46] Hermiticity of the lagrangian ensures probability conservation (unitarity condition). A "local" lagrangian is composed only of terms containing products of fields at the same space-time point.

$$\mathfrak{L}_i(t, \boldsymbol{x}) = g \left(\overline{\psi}_a \psi_b \right) \left(i \overline{\psi}_c \gamma_5 \psi_d \right) \xrightarrow{CPT} \mathfrak{L}_i^\dagger(-t, -\boldsymbol{x}), \tag{E.59}$$

where we used the fact that $\hat{T}$ implies charge conjugation: c number $\to$ (c number)*. For $\mathfrak{L}$ to be hermitian, it must also contain a term $\mathfrak{L}_i^\dagger$. The sum $\mathfrak{L}_i + \mathfrak{L}_i^\dagger$ is evidently CPT invariant.

The same holds true for a combination of vector and/or axial vector fields, e.g.,

$$\mathfrak{L}_j(t, \boldsymbol{x}) = g \, V^\mu(t, \boldsymbol{x}) A_\mu(t, \boldsymbol{x}) + h.c. \xrightarrow{CPT} \mathfrak{L}_j(-t, -\boldsymbol{x}), \tag{E.60}$$

where h.c. denotes hermitian conjugate. Since tensors transform as products of vectors and/or axial vectors, we conclude that

$$[CPT] \, \mathfrak{L}(x) \, [CPT]^{-1} = \mathfrak{L}(-x). \tag{E.61}$$

Ignoring irrelevant phases, transformation relations (E.14), (E.26) and (E.53) amount to

$$\boldsymbol{b}^\dagger(\boldsymbol{p}, s) \xrightarrow{P} \boldsymbol{b}^\dagger(-\boldsymbol{p}, s) \xrightarrow{C} \boldsymbol{d}^\dagger(-\boldsymbol{p}, s) \xrightarrow{T} \boldsymbol{d}^\dagger(\boldsymbol{p}, -s). \tag{E.62}$$

The combined CPT operation converts particles to antiparticles and exchanges kets and bras. Under this symmetry transformation, the momentum of the particle is unchanged because both space and time are reflected, but the sign of the spin state is reversed.

We will round off our discussion of the discrete symmetry transformations C, P and T by deriving the entries in Table E.2. Consider first the transformation properties of the Dirac bilinears $\overline{\psi}_a \Gamma_i \psi_b$ under CP:

$$\begin{aligned}
\overline{\psi}_a^{cp} \Gamma_i \psi_b^{cp} &= \boldsymbol{U}(C) \overline{\psi}_a^p \boldsymbol{U}^{-1}(C) \, \Gamma_i \boldsymbol{U}(C) \psi_b^p \boldsymbol{U}^{-1}(C) \\
&= \overline{\psi}_b(t, -\boldsymbol{x}) \Gamma_i^{cp} \psi_a(t, -\boldsymbol{x}),
\end{aligned} \tag{E.63}$$

where

$$\Gamma_i = \mathbb{1}, \gamma_5, \gamma^\mu, \gamma^\mu \gamma_5, \sigma^{\mu\nu} \equiv \frac{i}{2} \left(\gamma^\mu \gamma^\nu - \gamma^\nu \gamma^\mu \right) \tag{E.64}$$

($\mathbb{1}$ is the unit 4×4 matrix) and

$$\Gamma_i^{cp} \equiv \gamma^0 (\hat{C}^{-1} \Gamma_i \hat{C})^{\mathrm{T}} \gamma^0. \tag{E.65}$$

Expressions (E.63) and (E.65) were derived by using

$$\begin{aligned}
\boldsymbol{U}(C) \psi_b^p(x) \boldsymbol{U}^{-1}(C) &= \gamma^0 \boldsymbol{U}(C) \psi_b(t, -\boldsymbol{x}) \boldsymbol{U}^{-1}(C) \\
&= -\hat{C} \gamma^0 \overline{\psi}_b^{\mathrm{T}}(t, -\boldsymbol{x}) \\
\boldsymbol{U}(C) \overline{\psi}_a^p(x) \boldsymbol{U}^{-1}(C) &= \boldsymbol{U}(C) \overline{\psi}_a(t, -\boldsymbol{x}) \boldsymbol{U}^{-1}(C) \gamma^0 \\
&= \left[\boldsymbol{U}(C) \psi_a \boldsymbol{U}^{-1}(C) \right]^\dagger = \psi_a^{\mathrm{T}}(t, -\boldsymbol{x}) \gamma^0 \hat{C}^{-1}.
\end{aligned} \tag{E.66}$$

To evaluate Γ_i^{cp}, note that

$$
\begin{aligned}
\hat{C}^{-1}\gamma_5\hat{C} &= \gamma_5 = (\gamma_5)^{\mathrm{T}} \\
\hat{C}^{-1}\gamma^\mu\gamma^\nu\hat{C} &= \hat{C}^{-1}\gamma^\mu\hat{C}\hat{C}^{-1}\gamma^\nu\hat{C} = (\gamma^\mu)^{\mathrm{T}}(\gamma^\nu)^{\mathrm{T}} = -(\gamma^\mu\gamma^\nu)^{\mathrm{T}}, \\
\hat{C}^{-1}\gamma^\mu\gamma_5\hat{C} &= (-\gamma^\mu)^{\mathrm{T}}\hat{C}^{-1}\gamma_5\hat{C} = (-\gamma^\mu)^{\mathrm{T}}(\gamma_t)^{\mathrm{T}} = (\gamma^\mu\gamma_t)^{\mathrm{T}},
\end{aligned}
\tag{E.67}
$$

based on (E.20). It is then straightforward to show that

$$
\Gamma_i^{cp} = \mathbb{1},\quad -\gamma_5,\quad -\gamma_\mu,\quad -\gamma_\mu\gamma_5,\quad -\sigma_{\mu\nu},
\tag{E.68}
$$

where we set $\gamma^0\gamma^\mu\gamma^0 = \gamma_\mu$.

Under the time-reversal transformation,

$$
\begin{aligned}
\overline{\psi}_a^t\,\Gamma_i\,\psi_b^t &= \boldsymbol{V}(T)\,\overline{\psi}_a(x)\boldsymbol{V}^{-1}(T)\,\Gamma_i^*\,\boldsymbol{V}(T)\,\psi_b(x)\boldsymbol{V}^{-1}(T) \\
&= \overline{\psi}_a(-t,\boldsymbol{x})\,\Gamma_i^t\,\psi_b(-t,\boldsymbol{x}),
\end{aligned}
\tag{E.69}
$$

where

$$
\Gamma_i^t \equiv \left(\gamma^1\gamma^3\right)^\dagger \Gamma_i^* \gamma^1\gamma^3
\tag{E.70}
$$

and (see (E.48a))[47]

$$
\begin{aligned}
\boldsymbol{V}(T)\,\psi(x)\boldsymbol{V}^{-1}(T) &= \gamma^1\gamma^3\psi(-t,\boldsymbol{x}), \\
\boldsymbol{V}(T)\,\overline{\psi}(x)\boldsymbol{V}^{-1}(T) &= \boldsymbol{V}(T)\,\psi^\dagger(x)\boldsymbol{V}^{-1}\gamma^0 \\
&= \left[\boldsymbol{V}(T)\,\psi(x)\boldsymbol{V}^{-1}(T)\right]^\dagger \gamma^0 \\
&= \overline{\psi}(-t,\boldsymbol{x})\left(\gamma^1\gamma^3\right)^\dagger.
\end{aligned}
\tag{E.71}
$$

From (E.70) we obtain

$$
\Gamma_i^t = \mathbb{1}, \gamma_5, \gamma_\mu, \gamma_\mu\gamma_5, -\sigma_{\mu\nu}.
\tag{E.72}
$$

Finally, under the combined CPT transformation,

$$
\begin{aligned}
\overline{\psi}_a^{cpt}\,\Gamma_i\,\psi_b^{cpt} &= \boldsymbol{V}(T)\,\overline{\psi}_a^{cp}\,\boldsymbol{V}^{-1}(T)\,\Gamma_i^*\,\boldsymbol{V}(T)\,\psi_b^{cp}\,\boldsymbol{V}^{-1}(T) \\
&= \overline{\psi}_b(-x)\,\Gamma_i^{cpt}\,\psi_a(-x),
\end{aligned}
\tag{E.73}
$$

where

$$
\Gamma_i^{cpt} \equiv \left(\gamma_5\gamma^0\,\Gamma_i^*\,\gamma^0\gamma_5\right)^{\mathrm{T}} = \gamma_5\gamma^0\,\Gamma_i^\dagger\,\gamma^0\gamma_5.
\tag{E.74}
$$

To derive (E.73) and (E.74), we used

$$
\begin{aligned}
\boldsymbol{V}(T)\,\psi_b^{cp}\,\boldsymbol{V}^{-1}(T) &= \gamma^1\gamma^3\psi_b^{cp}(-t,\boldsymbol{x}) = \gamma^1\gamma^3\left(-\hat{C}\gamma^0\,\overline{\psi}_b^{\mathrm{T}}\right) \\
&= \gamma^0\gamma_5\overline{\psi}_b^{\mathrm{T}}(-x) \\
\boldsymbol{V}(T)\,\overline{\psi}_a^{cp}\boldsymbol{V}^{-1}(T) &= (\psi_a^{cp})^\dagger\left(\gamma^1\gamma^3\right)^\dagger\gamma^0 = \left(-\hat{C}\gamma^0\,\overline{\psi}_a^{\mathrm{T}}\right)^\dagger\left(-\gamma^1\gamma^3\gamma^0\right) \\
&= -\psi_a^{\mathrm{T}}(-x)\gamma_5\gamma^0.
\end{aligned}
\tag{E.75}
$$

---

[47] Note that $\boldsymbol{V}\psi^\dagger\boldsymbol{V}^{-1} = \boldsymbol{U}\psi^{\mathrm{T}}\boldsymbol{U}^{-1} = \left[\boldsymbol{U}\psi^*\boldsymbol{U}^{-1}\right]^\dagger = \left[\boldsymbol{V}\psi\boldsymbol{V}^{-1}\right]^\dagger.$

Upon substituting

$$(\gamma^\mu)^\dagger = \gamma^0\gamma^\mu\gamma^0 \tag{E.76}$$

in expression (E.74), it yields

$$\Gamma_i^{cpt} = \mathbb{1}, \quad -\gamma_5, \quad -\gamma^\mu, \quad -\gamma^\mu\gamma_5, \quad \sigma^{\mu\nu}. \tag{E.77}$$

In general, the transformation properties of a free Dirac field under the discrete symmetry operations P, C and T can be expressed as

$$\begin{aligned}
\boldsymbol{U}(P)\psi(x)\boldsymbol{U}^{-1}(P) &= \eta_p\gamma^0\psi(t,-\boldsymbol{x}), \\
\boldsymbol{U}(P)\overline{\psi}(x)\boldsymbol{U}^{-1}(P) &= \eta_p^*\,\overline{\psi}(t,-\boldsymbol{x})\gamma^0, \\
\boldsymbol{U}(C)\psi(x)\boldsymbol{U}^{-1}(C) &= \eta_c\hat{C}\overline{\psi}^{\mathrm{T}}(t,\boldsymbol{x}), \\
\boldsymbol{U}(C)\overline{\psi}(x)\boldsymbol{U}^{-1}(C) &= -\eta_c^*\,\psi^{\mathrm{T}}(x)\hat{C}^{-1}, \\
\boldsymbol{V}(T)\psi(x)\boldsymbol{V}^{-1}(T) &= \eta_t\hat{T}\psi(-t,\boldsymbol{x}), \\
\boldsymbol{V}(T)\overline{\psi}(x)\boldsymbol{V}^{-1}(T) &= \eta_t^*\,\overline{\psi}(-t,\boldsymbol{x})\hat{T}^\dagger,
\end{aligned} \tag{E.78}$$

where η_p, η_c and η_t are arbitrary phases. One can easily show that these phases do not enter into the transformation laws for the Dirac bilinears.

F The Vacuum Insertion Approximation

If we define $O_\mu \equiv \gamma_\mu(1-\gamma_5)$, the matrix element of the four-fermion operator in (9.17) reads

$$\begin{aligned}
\mathcal{M} &\equiv \langle \bar{K}^0|\,\bar{s}O_\mu d\,\bar{s}O^\mu d\,|K^0\rangle \\
&= O_{ij}O_{k\ell}\,\langle \bar{K}^0|\,\bar{s}^{i\alpha}\delta_{\alpha\sigma}d^{j\sigma}\bar{s}^{k\gamma}\delta_{\gamma\tau}d^{\ell\tau}\,|K^0\rangle,
\end{aligned} \tag{F.1}$$

where Greek letters denote the color indices (1,2,3) of quark creation and annihilation operators, and $\bar{s}Od \equiv \sum_\alpha \bar{s}^\alpha Od^\alpha$. Like weak quark currents, physical hadrons are colorless; that is, they are singlets in color space.

In order to estimate the magnitude of $\mathcal{M}$, it is customary to insert the vacuum state between the two $\bar{s}Od$ currents. In this approximation, the operators are treated as free fields (i.e., strong interactions are neglected) and the K^0 ($\bar{K}^0$) is considered as made up of only $\bar{s}d$ ($\bar{d}s$). Taking into account that quark fields anticommute, one can form the following combinations of two-fermion operators:

$$\begin{aligned}
\hat{\mathcal{O}} = O_{ij}O_{k\ell}\big\{ &\bar{s}^{i\alpha}d^{j\alpha}\cdot\bar{s}^{k\gamma}d^{\ell\gamma} + \bar{s}^{k\gamma}d^{\ell\gamma}\cdot\bar{s}^{i\alpha}d^{j\alpha} \\
&- \bar{s}^{i\alpha}\delta_{\alpha\sigma}d^{\ell\tau}\cdot\bar{s}^{k\gamma}\delta_{\gamma\tau}d^{j\sigma} - \bar{s}^{k\gamma}\delta_{\gamma\tau}d^{j\sigma}\cdot\bar{s}^{i\alpha}\delta_{\alpha\sigma}d^{\ell\tau}\big\}.
\end{aligned} \tag{F.2}$$

To simplify the calculation, we use the Fierz identities

$$O_{ij}O_{k\ell} = -O_{i\ell}O_{kj}, \quad \delta_{\alpha\sigma}\delta_{\gamma\tau} = \frac{1}{3}\delta_{\alpha\tau}\delta_{\gamma\sigma} + \frac{1}{2}\lambda_{\alpha\tau}^a\lambda_{\gamma\sigma}^a, \tag{F.3}$$

where λ^a are the 3×3 color generator matrices ($a = 1, \ \dots \ ,8$). If the two-fermion operators are inserted between the vacuum and the colour singlet K^0 state, the octet–octet term cannot contribute. Therefore,

$$\mathcal{M} = 2 \langle \bar{K}^0 | \, \bar{s} O_\mu d \, | \, 0 \rangle \langle 0 \, | \, \bar{s} O^\mu d \, | K^0 \rangle$$

$$+ \frac{1}{3} \langle \bar{K}^0 | \, \bar{s}^{i\alpha} O_{i\ell} d^{\ell\alpha} \, | \, 0 \rangle \langle 0 \, | \, \bar{s}^{k\gamma} O_{kj} d^{j\gamma} \, | K^0 \rangle$$

$$+ \frac{1}{3} \langle \bar{K}^0 | \, \bar{s}^{k\gamma} O_{kj} d^{j\gamma} \, | \, 0 \rangle \langle 0 \, | \, \bar{s}^{i\alpha} O_{i\ell} d^{\ell\alpha} \, | K^0 \rangle$$

$$= 2 \left(1 + \frac{1}{3} \right) | \langle \bar{K}^0 | \, \bar{s} O d \, | K^0 \rangle |^2$$

$$= \frac{8}{3} (f_k m_k)^2. \tag{F.4}$$

References

1. G. Rochester and C. Butler, Nature **160**, 855 (1947).
2. M. Gell-Mann, Phys. Rev. **92**, 833 (1953);
   T. Nakano and K. Nishijima, Prog. Theor. Phys. **10**, 581 (1953).
3. M. Gell-Mann and A. Pais, Phys. Rev. **97**, 1387 (1955).
4. K. Lande et al., Phys. Rev. **103**, 1901 (1956).
5. K. Lande et al., Phys. Rev. **105**, 1925 (1957).
6. E. Boldt et al., Phys. Rev. Lett. **1**, 150 (1958).
7. U. Camerini et al., Phys. Rev. **128**, 362 (1962).
8. A. Pais and O. Piccioni, Phys. Rev. **100**, 1487 (1955);
   M. L. Good, Phys. Rev. **106**, 591 (1957).
9. F. Muller et al., Phys. Rev. Lett. **4**, 418 (1960);
   R. H. Good et al., Phys. Rev. **124**, 1223 (1961).
10. J. Christenson et al., Phys.Rev. **140**, B74 (1965).
11. T. Fujii et al., Phys. Rev. Lett. **13**, 253 (1964).
12. W. Mehlhop et al., Phys. Rev. **172**, 1613 (1968).
13a. J. Christenson et al., Phys. Rev. Lett. **13**, 138 (1964);
13b. J.-M. Gaillard et al., Phys. Rev. Lett. **18**, 20 (1967);
   J. Cronin et al., Phys. Rev. Lett. **18**, 25 (1967);
13c. E. Ramberg et al., Phys. Rev. Lett. **70**, 2525 and 2529 (1993).
14a. J. Bell and J. Steinberger, Proc. 1965 Oxford Intl. Conf. on Elementar Particles;
14b. J. Steinberger, in *CP Violation in Part. Physics and Astrophysics*, (Blois, 1989);
14c. J. Bell, 1965 Intl. School of Physics "Etore Majorana" (Academic Press, New York, 1966);
14d. T. T. Wu and C. N. Yang, Phys. Rev. Lett. **13**, 380 (1964).
15. B. Aubert et al., Phys. Lett. **17**, 59 (1965).
16. E. Stückelberg, Helv. Phys. Acta **25**, 577 (1952).
17. V. Fitch et al., Phys. Rev. Lett. **15**, 73 (1965) and Phys. Rev. **164**, 1711 (1967).
18. C. Alff-Steinberger et al., Phys. Lett. **20**, 207 (1966); **21**, 595 (1966).
19. M. Bott-Bodenhausen et al., Phys. Lett. **20**, 212 (1966); **23**, 277 (1966).
20. J.-M. Gaillard, *Methods in Subnucl. Physics I*, ed. by M. Nikolić, (Gordon and Breach, New York, 1968).
21. Review of Particle Physics, Phys. Rev. D **54**, 1 (1996).
22. F. Niebergall et al., Phys. Lett. B **49**, 103 (1974).
23. S. Gjesdal et al., Phys. Lett. B **52**, 113 (1974).
24. C. Geweniger et al., Phys. Lett. B **48**, 483 (1974).
25. W. Carithers et al., Phys. Rev. Lett. **34**, 1240 and 1244 (1975).
26. S. Bennet et al., Phys. Rev. Lett. **19**, 993 (1967).
27. D. Dorfan et al., Phys. Rev. Lett. **19**, 987 (1967).

28. C. Geweniger et al., Phys. Lett. B **52**, 108 (1974).
29. C. Geweniger et al., Phys. Lett. B **48**, 487 (1974).
30. S. Gjesdal et al., Phys. Lett. B **52**, 119 (1974).
31a. H. Burkhardt et al., Phys. Lett. B **206**, 169 (1988);
31b. G. Barr et al., Phys. Lett. B **317**, 233 (1993).
32. R. Carosi et al., Phys. Lett. B **237**, 303 (1990).
33. M. Woods et al., Phys. Rev. Lett. **60**, 1695 (1988).
34. J. Roehrig et al., Phys. Rev. Lett. **38**, 1116 (1977);
    R. Briere and B. Winstein, Phys. Rev. Lett. **75**, 402 (1995).
35. L. Gibbons et al., Phys. Rev. Lett. **70**, 1199 and 1203 (1993).
36. B. Schwingenheuer et al., Phys. Rev. Lett. **74**, 4376 (1995).
37. B. Winstein and L. Wolfenstein, Rev. Mod. Phys. **74**, 1113 (1993).
38. R. Armenteros et al., Proc. Intl. Conf. on HEP, Geneva, 1962;
    P. Franzini et al., Phys. Rev. **140**, B127 (1965).
39a. R. Adler et al., Phys. Lett. B **363**, 243 (1995);
39b. R. Adler et al., Phys. Lett. B **369**, 367 (1996).
40. R. Adler et al., Phys. Lett. B **363**, 237 (1995).
41. (a),(b) A. Angelopoulos et al., Phys. Lett. B **444**, 38(a), 43(b) (1988).
42a. R. Adler et al., CERN-PPE/97-54 (1997).
42b. R. Adler et al., Phys. Lett. B **370**, 167 (1996).
43. Y. Zou et al., Phys. Lett. B **329**, 519 (1994).
44. A. Aharony, Lett. Nuovo Cimento **3**, 791 (1970);
    P. Kabir, Phys. Rev. D **2**, 540 (1970).
45. G. Wick, Ann. Rev. Nucl. Sci. **9**, 1 (1958).
46a. H. Lipkin, Phys. Rev. **176**, 1715 (1968);
46b. I. Dunietz, J. Hauser and J. Rosner, Phys. Rev. D **35**, 2166 (1987);
46c. C. Buchanan et al., Phys. Rev. D **45**, 4088 (1992).
47. F. Selleri, Phys. Rev. A **56**, 3493 (1997).
48. A. Apostolakis et al., CERN-PPE/97-140 (1997).
49. P. Grafström et al., NIM A **344**, 487 (1994).
50. S. Glashow, J. Iliopoulos and L Maiani, Phys. Rev. D **2**, 1285 (1970).
51a. B. Lee, J. Primack and S. Treiman, Phys. Rev. D **7**, 510 (1973).
51b. M. Gaillard and B. Lee, Phys. Rev. D **10**, 897 (1974).
52. M. Kobayashi and K. Maskawa, Prog. Theor. Phys. **49**, 652 (1973).
53. L. Wolfenstein, Phys. Rev. Lett. **51**, 1945 (1983).
54. G. Buchalla, A. Buras and M. Lautenbacher, Rev. Mod. Phys. **68**, 1125 (1996).
55. T. Inami and C. Lim, Prog. Theor. Phys. **65**, 297 (1981).
56. S. Stone, HEPSY 96-01 (1996).
57. S. L. Wu, CERN-PPE/96-82 (1996).
58. DELPHI Collaboration, CERN-PPE/97-114 (1997).
59. L.-L. Chau and W.-Y. Keung, Phys. Rev. Lett. **53**, 1802 (1984).
    C. Jarlskog, Phys. Rev. Lett. **55**, 1039 (1985).
60. C. Albajar et al., Phys. Lett. B **186**, 247 (1987).
61. H. Albrecht et al., Phys. Lett. B **192**, 245 (1987).
62. L. Littenberg, Phys. Rev. D **39**, 3322 (1989).
63. W. Marciano and Z. Parsa, Phys. Rev. D **53**, R1 (1996).
64. M. Weaver et al., Phys. Rev. Lett. **72**, 3758 (1994).
65. K. Arisaka et al., Fermilab FN-568 (1991); The KAMI Collab., EOI (1997).
66. T. Inagaki et al., KEK Internal Report 96-13 (1996).
67. I.-H. Chiang et al., BNL Proposal P926 (1996).
68. D. Rein and L. Sehgal, Phys. Rev. D **39**, 3325 (1989).
69. G. Buchalla and A. Buras, Nucl. Phys. B **412**, 106 (1994).

70. M. Atiya et al., Nucl. Instr. and Methods A **321**, 129 (1992).
71. S. Adler et al., Phys. Rev. Lett. **79**, 2204 (1997).
72. W. Marciano, in *Rare Decay Symposium* (World Scientific, Singapore, 1989).
73. L. Sehgal, Phys. Rev. **183**, 1511 (1969);
    B. Martin, E. de Rafael and J. Smith, Phys. Rev. D **2**, 179 (1970).
74. D. Ambrose et al., Phys. Rev. Lett. **81**, 4309 (1998).
75. A. Heinson et al., Phys. Rev. D **51**, 985 (1995);
    T. Akagi et al., Phys. Rev. D **51**, 2061 (1995);
76. P. Herczeg, Phys. Rev. D **27**, 1512 (1983).
77. G. Ecker and A. Pich, Nucl. Phys. B **366**, 189 (1991).
78. F. Gilman and M. Wise, Phys. Lett. B **83**, 83 (1979).
79. J. Bijnens and M. Wise, Phys. Lett. B **137**, 245 (1984).
80. H. Braun et al., Nucl. Phys. B **89**, 210 (1975).
81. N. Cabibbo and A. Maksymowicz, Phys. Lett. **9**, 352 (1964).
82. Lee T. D., Oehme R. and Yang C. N., Phys. Rev. **106**, 340 (1957).
83. Sakurai J. J., *Invariance Principles and Elementary Particles* (Princeton University Press, Princeton, NJ, 1964).
84. Greiner W. and Müller B., *Gauge Theory of Weak Interactions*, 2nd edn. (Springer, Heidelberg, Berlin, 1996).
85. Lee T. D. and Wu C. S., Ann. Rev. Nucl. Sci. **16**, 511 (1966).
86. Charpak G. and Gourdin M., The $K^0\bar{K}^0$ system, CERN Yellow Report 67-18 (1967).
87. Kabir P., *The CP Puzzle* (Academic Press, New York, 1968).
88. Faissner H., in *Lectures in Theoretical Physics*, ed. by K. Mahanthappa, W. Brittin and A. Barut (Gordon and Breach, New York, 1969).
89. Kleinknecht K., K_L-K_S Regeneration, Fortschritte der Physik **21**, 57 (1973).
90. Gaillard J.-M., in Weak Interactions, School of Elementary Particle Phys., Baško Polje, 1973.
91. Kleinknecht K., Ann. Rev. Nucl. Sci. **26**, 1 (1976).
92. Cronin J., "CP symmetry violation — the search for its origin", Rev. Mod. Phys. **53**, 373 (1981).
93. Fitch V., "The discovery of charge-conjugation parity asymmetry", Rev. Mod. Phys. **53**, 367 (1981).
94. Okun L., *Leptons and Quarks* (North-Holland, Amsterdam, 1982).
95. Commins E. and Bucksbaum P., *Weak Interactions of Leptons and Quarks* (Cambridge University Press, Cambridge, 1983).
96. Tanner N. and Dalitz R. H., Annals of Physics **171**, 463 (1986).
97. Sachs R., *The Physics of Time Reversal* (University of Chicago Press, Chicago, IL, 1987).
98. Altarelli G., "Three lectures on flavour mixing", in *Techniques and Concepts of High-Energy Physics* (Plenum, New York, 1988).
99. Peccei R., in The 1988 Theor. Adv. Study Inst. in Elem. Part. Physics (TASI-88), Brown University, World Scientific, 1989.
100. Li L.-F., "Rare kaon decays", in *Quarks, Mesons and Nuclei* (World Scientific, Singapore, 1989).
101. Wolfenstein L. (Ed.), *CP Violation* (North-Holland, Amsterdam, 1989).
102. Jarlskog C., "Introduction to CP violation", in *CP Violation*, ed. by C. Jarlskog (World Scientific, Singapore, 1989).
103. Nachtmann O., *Elementary Particle Physics* (Springer, Heidelberg, Berlin, 1990).
104. Winstein B., "Topics in kaon physics", in *Techniques and Concepts of High-Energy Physics*, ed. by T. Ferbel (Plenum, New York, 1990).
105. Nelson H. N., in SLAC Summer Institute on Particle Physics, 1992.

106. Donoghue J., Golowich E. and Holstein B., *Dynamics of the Standard Model* (Cambridge University Press, Cambridge, 1992).
107. Ritchie J. and Wojcicki S., "Rare K decays", Rev. Mod. Phys. **65**, 1149 (1993).
108. Leader E. and Predazzi E., *An Introduction to Gauge Theories and Modern Particle Physics* (Cambridge University Press, Cambridge, 1996).
109. Buras A., Weak Hamiltonian, CP Violation and Rare Decays, TUM-HEP-316/98 (1998).

Name Index

Subject Index

Subject Index (Decays)

Computer to plate: Mercedes Druck, Berlin
Binding: Buchbinderei Lüderitz & Bauer, Berlin

Springer Tracts in Modern Physics